T0187431

Second Edition

Gauge Fields

Introduction to Quantum Theory

L. D. Faddeev
Academy of Sciences U.S.S.R.
Steclov Mathematical Institute, Leningrad

and

A. A. Slavnov
Academy of Sciences U.S.S.R.
Steclov Mathematical Institute, Moscow

Translated from the Russion Edition
by G. B. Pontecorvo
Joint Institute For Nuclear Research, Dubna

CRC Press
Taylor & Francis Group
Boca Raton London New York

CRC Press is an imprint of the
Taylor & Francis Group, an Informa business

First published 1991 by Westview Press

Published 2018 by CRC Press
Taylor & Francis Group
6000 Broken Sound Parkway NW, Suite 300
Boca Raton, FL 33487-2742

ISBN 13: 978-0-201-40634-4 (pbk)

Visit the Taylor & Francis Web site at
http://www.taylorandfrancis.com

and the CRC Press Web site at
http://www.crcpress.com

Library of Congress Cataloging-in-Publication Data

Slavnov, A. A. (Andrei Alekseevich)
[Vvedenie v kvantovuiu teoriiu kalibrovochnykh polei. English]
Gauge fields, introduction to quantum theory / L.D. Faddeev,
A.A. Slavnov; translated from the second Russian edition by
G.B. Pontecorvo.
p. cm.
Translation of: Vvendenie v kvantovuiu teoriiu kalibrovochnykh
polei / A.A. Slavnov, L.D. Faddeev. Izd. 2., perer. i dop. 1988.
1. Gauge fields (Physics) 2. Quantum field theory. I. Faddeev,
L.D. II. Title.
QC793.3.F5S5313 1991 91-846
530.1'435--dc20
ISBN 0-201-52472-4

Gauge Fields

Introduction to Quantum Theory

Frontiers in Physics

DAVID PINES/Editor

Volumes of the Series published from 1961 to 1973 are not officially numbered. The parenthetical numbers shown are designed to aid librarians and bibliographers to check the completeness of their holdings.

Titles published in this series prior to 1987 appear under either the W. A. Benjamin or the Benjamin/Cummings imprint; titles published since 1986 appear under the Addison-Wesley imprint.

(1)	N. Bloembergen	Nuclear Magnetic Relaxation: A Reprint Volume, 1961
(2)	G. F. Chew	S-Matrix Theory of Strong Interactions: A Lecture Note and Reprint Volume, 1961
(3)	R. P. Feynman	Quantum Electrodynamics: A Lecture Note and Reprint Volume, 1961
(4)	R. P. Feynman	The Theory of Fundamental Processes: A Lecture Note Volume, 1961
(5)	L. Van Hove, N. M. Hugenholtz and L. P. Howland	Problem in Quantum Theory of Many-Particle Systems: A Lecture Note and Reprint Volume, 1961
(6)	D. Pines	The Many-Body Problem: A Lecture Note and Reprint Volume, 1961
(7)	H. Frauenfelder	The Mössbauer Effect: A Review—With a Collection of Reprints, 1962
(8)	L. P. Kadanoff G. Baym	Quantum Statistical Mechanics: Green's Function Methods in Equilibrium and Nonequilibrium Problems, 1962
(9)	G. E. Pake	Paramagnetic Resonance: An Introductory Monograph, 1962 [cr. (42)—2nd edition]
(10)	P. W. Anderson	Concepts in Solids: Lectures on the Theory of Solids, 1963
(11)	S. C. Frautschi	Regge Poles and S-Matrix Theory, 1963
(12)	R. Hofstadter	Electron Scattering and Nuclear and Nucleon Structure: A Collection of Reprints with an Introduction, 1963
(13)	A. M. Lane	Nuclear Theory: Pairing Force Correlations to Collective Motion, 1964
(14)	R. Omnès M. Froissart	Mandelstam Theory and Regge Poles: An Introduction for Experimentalists, 1963
(15)	E. J. Squires	Complex Angular Momenta and Particle Physics: A Lecture Note and Reprint Volume, 1963
(16)	H. L. Frisch J. L. Lebowitz	The Equilibrium Theory of Classical Fluids: A Lecture Note and Reprint Volume, 1964
(17)	M. Gell-Mann Y. Ne'eman	The Eightfold Way (A Review—With a Collection of Reprints), 1964
(18)	M. Jacob G. F. Chew	Strong-Interaction Physics: A Lecture Note Volume, 1964
(19)	P. Nozières	Theory of Interacting Fermi Systems, 1964
(20)	J. R. Schrieffer	Theory of Superconductivity, 1964 (revised 3rd printing, 1983)
(21)	N. Bloembergen	Nonlinear Optics: A Lecture Note and Reprint Volume, 1965
(22)	R. Brout	Phase Transitions, 1965
(23)	I. M. Khalatnikov	An Introduction to the Theory of Superfluidity, 1965
(24)	P. G. deGennes	Superconductivity of Metals and Alloys, 1966
(25)	W. A. Harrison	Pseudopotentials in the Theory of Metals, 1966
(26)	V. Barger D. Cline	Phenomenological Theories of High Energy Scattering: An Experimental Evaluation, 1967
(27)	P. Choquàrd	The Anharmonic Crystal, 1967

(28)	T. Loucks	Augmented Plane Wave Method: A Guide to Performing Electronic Structure Calculations—A Lecture Note and Reprint Volume, 1967
(29)	Y. Ne'eman	Algebraic Theory of Particle Physics: Hadron Dynamics in Terms of Unitary Spin Current, 1967
(30)	S. L. Adler R. F. Dashen	Current Algebras and Applications to Particle Physics, 1968
(31)	A. B. Migdal	Nuclear Theory: The Quasiparticle Method, 1968
(32)	J. J. J. Kokkedee	The Quark Model, 1969
(33)	A. B. Migdal V. Krainov	Approximation Methods in Quantum Mechanics, 1969
(34)	R. Z. Sagdeev and A. A. Galeev	Nonlinear Plasma Theory, 1969
(35)	J. Schwinger	Quantum Kinematics and Dynamics, 1970
(36)	R. P. Feynman	Statistical Mechanics: A Set of Lectures, 1972
(37)	R. P. Feynman	Photon-Hadron Interactions, 1972
(38)	E. R. Caianiello	Combinatorics and Renormalization in Quantum Field Theory, 1973
(39)	G. B. Field, H. Arp, and J. N. Bahcall	The Redshift Controversy, 1973
(40)	D. Horn F. Zachariasen	Hadron Physics at Very High Energies, 1973
(41)	S. Ichimaru	Basic Principles of Plasma Physics: A Statistical Approach, 1973 (2nd printing, with revisions, 1980)
(42)	G. E. Pake T. L. Estle	The Physical Principles of Electron Paramagnetic Resonance, 2nd Edition, completely revised, enlarged, and reset, 1973 [cf. (9)—1st edition]

Volumes published from 1974 onward are being numbered as an integral part of the bibliography.

43	R. C. Davidson	Theory of Nonneutral Plasmas, 1974
44	S. Doniach E. H. Sondheimer	Green's Functions for Solid State Physicists, 1974
45	P. H. Frampton	Dual Resonance Models, 1974
46	S. K. Ma	Modern Theory of Critical Phenomena, 1976
47	D. Forster	Hydrodynamic Fluctuations, Broken Symmetry, and Correlation Functions, 1975
48	A. B. Migdal	Qualitative Methods in Quantum Theory, 1977
49	S. W. Lovesey	Condensed Matter Physics: Dynamic Correlations, 1980
50	L. D. Faddeev A. A. Slavnov	Gauge Fields: Introduction to Quantum Theory, 1980
51	P. Ramond	Field Theory: A Modern Primer, 1981 [cf. 74—2nd ed.]
52	R. A. Broglia A. Winther	Heavy Ion Reactions: Lecture Notes Vol. I, Elastic and Inelastic Reactions, 1981
53	R. A. Broglia A. Winther	Heavy Ion Reactions: Lecture Notes Vol. II, 1990
54	H. Georgi	Lie Algebras in Particle Physics: From Isospin to Unified Theories, 1982
55	P. W. Anderson	Basic Notions of Condensed Matter Physics, 1983
56	C. Quigg	Gauge Theories of the Strong, Weak, and Electromagnetic Interactions, 1983
57	S. I. Pekar	Crystal Optics and Additional Light Waves, 1983

58	S. J. Gates M. T. Grisaru M. Roček W. Siegel	Superspace *or* One Thousand and One Lessons in Supersymmetry, 1983
59	R. N. Cahn	Semi-Simple Lie Algebras and Their Representations, 1984
60	G. G. Ross	Grand Unified Theories, 1984
61	S. W. Lovesey	Condensed Matter Physics: Dynamic Correlations, 2nd Edition, 1986
62	P. H. Frampton	Gauge Field Theories, 1986
63	J. I. Katz	High Energy Astrophysics, 1987
64	T. J. Ferbel	Experimental Techniques in High Energy Physics, 1987
65	T. Applequist A. Chodos P. G. O. Freund	Modern Kaluza-Klein Theories, 1987
66	G. Parisi	Statistical Field Theory, 1988
67	R. C. Richardson E. N. Smith	Techniques in Low-Temperature Condensed Matter Physics, 1988
68	J. W. Negele H. Orland	Quantum Many-Particle Systems, 1987
69	E. W. Kolb M. S. Turner	The Early Universe, 1990
70	E. W. Kolb M. S. Turner	The Early Universe: Reprints, 1988
71	V. Barger R. J. N. Phillips	Collider Physics, 1987
72	T. Tajima	Computational Plasma Physics, 1989
73	W. Kruer	The Physics of Laser Plasma Interactions, 1988
74	P. Ramond	Field Theory: A Modern Primer 2nd edition, 1989 [cf. 51—1st edition]
75	B. F. Hatfield	Quantum Field Theory of Point Particles and Strings, 1989
76	P. Sokolsky	Introduction to Ultrahigh Energy Cosmic Ray Physics, 1989
77	R. Field	Applications of Perturbative QCD, 1989
80	J. F. Gunion H. E. Haber G. Kane S. Dawson	The Higgs Hunter's Guide, 1990
81	R. C. Davidson	Physics of Nonneutral Plasmas, 1990
82	E. Fradkin	Field Theories of Condensed Matter Systems, 1991
83	L. D. Faddeev A. A. Slavnov	Gauge Fields, 1990
84	R. Broglia A. Winther	Heavy Ion Reactions, Parts I and II, 1990
85	N.Goldenfeld	Lectures on Phase Transitions and the Renormalization Group, 1991
86	R. D. Hazeltine J. D. Meiss	Plasma Confinement, 1991

EDITOR'S FOREWORD

The problem of communicating in a coherent fashion recent developments in the most exciting and active fields of physics continues to be with us. The enormous growth in the number of physicists has tended to make the familiar channels of communication considerably less effective. It has become increasingly difficult for experts in a given field to keep up with the current literature; the novice can only be confused. What is needed is both a consistent account of a field and the presentation of a definite "point of view" concerning it. Formal monographs cannot meet such a need in a rapidly developing field, while the review article seems to have fallen into disfavor. Indeed, it would seem that the people most actively engaged in developing a given field are the people least likely to write at length about it.

FRONTIERS IN PHYSICS was conceived in 1961 in an effort to improve the situation in several ways. Leading physicists frequently give a series of lectures, a graduate seminar, or a graduate course in their special fields of interest. Such lectures serve to summarize the present status of a rapidly developing field and may well constitute the only coherent account available at the time. Often, notes on lectures exist (prepared by the lecturer himself, by graduate students, or by postdoctoral fellows) and are distributed on a limited basis. One of the principal purposes of the FRONTIERS IN PHYSICS Series is to make such notes available to a wider audience of physicists.

It should be emphasized that lecture notes are necessarily rough and informal, both in style and content; and those in the series will prove no exception. This is as it should be. The point of the series is to offer new, rapid, more informal, and, it is hoped, more effective ways for physicists to teach one another. The point is lost if only elegant notes qualify.

The informal monograph, representing an intermediate step between lecture notes and formal monographs, offers an author the opportunity to present his views of a field which has developed to the point where a summation might prove extraordinarily fruitful but a formal monograph might be feasible or desirable.

During the past decade, the informal text monograph, *Gauge Fields,* has provided the reader with a lucid introduction to the role played by gauge fields in quantum field theory. As its eminent authors note, over the same period gauge invariant models have evolved from providing an attractive physical hypothesis into a working theory which describes accurately the physics of elementary particles at moderate energies. A second edition which contains both supplementary and improved material is therefore both timely and highly useful, and it gives me pleasure to welcome once more Drs. Faddeev and Slavnov to FRONTIERS IN PHYSICS.

DAVID PINES
Urbana, Illinois
September, 1990

Preface to the Second Revised (Russian) Edition

During the past ten years, since the first edition of this book, gauge invariant models of elementary particle interactions were transformed from an attractive plausible hypothesis into a generally accepted theory confirmed by experiments. It was therefore natural that the development of the methods of gauge fields attracted the attention of the great majority of specialists in quantum field theory. The new interesting lines of activity that arose in this period included the formulation of gauge theories on a lattice, the investigation of non-trivial classical solutions of the Yang-Mills equations and quantization in their neighborhood, the application of methods of algebraic topology in gauge field theory. In preparing the second edition of our book we were confronted with a difficult dilemma: either we were to extend the book significantly by including a serious discussion of the novel fields of research, or we would, in the main, adopt the same plan as for the first edition. We decided in favour of the latter version, since, in our opinion, the most promising issues mentioned above have not as yet attained a completed form. Besides, an exposition of these issues would require a significant extension of the mathematical apparatus utilized. Therefore, in the second edition we limited ourselves to presenting such supplements that are related in a natural way to the main content of the first edition, and we also introduced a number of improvements which, as we hope, should facilitate reading of the book and render it more self-consistent.

This Preface is being written just at a time, when hopes are arising that a more fundamental basis is to be developed for elementary particle theory, the theory of superstrings.

However, independently of whether these hopes come true, gauge field theory, clearly, describes the physics of elementary particles adequately at moderate energies. Besides, the methods applied in the field theory of relativistic strings represent a direct generalization of the methods of gauge field theory, to which this book is devoted. For this reason we consider a new edition of it to be useful, both for direct applications of the already developed gauge theory and for search of new ways.

Moscow - Leningrad, 1986

L. D. Faddeev
A. A. Slavnov

Preface to the Original (Russian) Edition

Progress in quantum field theory, during the last ten years, is to a great extent due to the development of the theory of Yang-Mills fields, sometimes called gauge fields. These fields open up new possibilities for the description of interactions of elementary particles in the framework of quantum field theory. Gauge fields are involved in most modern models of strong and also of weak and electromagnetic interactions. There also arise the extremely attractive prospects of unification of all the interactions into a single universal interaction.

At the same time the Yang-Mills fields have surely not been sufficiently considered in modern monographical literature. Although the Yang-Mills theory seems to be a rather special model from the of view of general quantum field theory, it is extremely specific and the models used in this theory are quite far from being traditional. The existing monograph of Konoplyova and Popov, "Gauge Fields", deals mainly with the geometrical aspects of the gauge field theory and illuminates the quantum theory of the Yang-Mills fields insufficiently. We hope that the present book to some extent will close this gap.

The main technical method, used in the quantum theory of gauge fields, is the path-integral method. Therefore, much attention is paid in this book to the description of this alternative approach to the quantum field theory. We have made an attempt to expound this method in a sufficiently self-consistent manner, proceeding from the fundamentals of quantum theory. Nevertheless, for a deeper understanding of the book it is desirable for the reader to be familiar with the traditional methods of quantum theory, for example, in the volume of the first four chapters of the book by N. N. Bogolubov and D. V. Shirkov, "Introduction to the Theory of Quantized Fields". In particular, we shall not go into details of comparing the Feynman diagrams to the terms of the perturbation-theory expansion, and of the rigorous substantiation of the renormalization procedure, based on the R-operation. These problems are not specific for the Yang-Mills theory and are presented in detail in the quoted monograph.

There are many publications on the Yang-Mills fields, and we shall not go into a detailed review of this literature to any extent. Our aim is to introduce the methods of the quantum Yang-Mills theory to the reader. We shall not discuss alternative approaches to this theory, but shall present in detail that approach, which seems to us the most simple and natural one. The applications dealt with in the book are illustrative in character and are not the last work to be said about applications of the Yang-Mills field to elementary-particle models. We do this consciously, since the phenomenological aspects of gauge theories are developing and changing rapidly. At the same time the technique of quantization and renormalization of the Yang-Mills fields has already become well established. Our book is mainly dedicated to these specific problems.

We are grateful to our colleagues of the V. A. Steclov Mathematical Institute in Moscow and Leningrad for numerous helpful discussions of the problems dealt with in this book.

We would especially like to thank D. V. Shirkov and O.I. Zav'ylov who read the manuscript and made many useful comments and E. Sh. Yegoryan for help in checking the formulas.

Moscow-Leningrad-Kirovsk L. D. Faddeev, A. A. Slavnov

Contents

1

Introduction: Fundamentals of Classical Gauge Field Theory

1.1 Basic Concepts and Notation

The theory of gauge fields at present represents the widely accepted theoretical basis of elementary particle physics. Indeed, the most elaborate model of field theory, quantum electrodynamics, is a particular case of the gauge theory. Further, models of weak interactions have acquired an elegant and self-consistent formulation in the framework of gauge theories. The phenomenological four-fermion interaction has been replaced by the interaction with an intermediate vector particle, the quantum of the Yang-Mills field. Existing experimental data along with the requirement of gauge invariance led to the prediction of weak neutral currents and of new quantum numbers for hadrons.

Phenomenological quark models of strong interactions also have their most natural foundation in the framwork of a gauge theory known as quantum chromodynamics. This theory provides a unique possibility of describing, in the framework of quantum field theory, the phenomenon of asymptotic freedom. This theory also affords hopes of explaining quark confinement, although this question is not quite clear.

Finally, the extension of the gauge principle may lead to the gravitational interaction also being placed in the general scheme of Yang-Mills fields.

So the possibility arises of explaining, on the basis of one principle, all the hierarchy of interactions existing in nature. The term unified field theory, discredited sometime ago, now acquires a new reality in the framework of gauge field theories. In the formation of this picture a number of scientists took part. Let us mention some of the key dates.

In 1953 C. N. Yang and R. L. Mills, for the first time, generalized the principle of gauge invariance of the interaction of electric charges to the case of interacting isospins. In their paper, they introduced a vector field, which later became known as the Yang-Mills field, and within the framework of the classical field theory its dynamics was developed.

In 1967 L. D. Faddeev, V. N. Popov, and B. De Witt constructed a self-consistent scheme for the quantization of massless Yang-Mills fields. In the same year, S. Weinberg and A. Salam independently proposed a unified gauge model ofweak and electromagnetic interactions, in which the electromagnetic field and the field of the intermediate vector boson were combined into a multiplet of Yang-Mills fields. This model was based on the mechanism of mass generation for vector bosons as a result of a spontaneous symmetry breaking, proposed earlier by P. Higgs and T. Kibble.

In 1971 G. t'Hooft showed that the general methods of quantization of massless Yang-Mills fields may be applied, practically without any change, to the case of spontaneously broken symmetry. Thus the possibility was discovered of constructing a self-consistent quantum theory of massive vector fields, which are necessary for the theory of weak interactions and, in particular, for the Salam-Weinberg model.

By 1972 the construction of the quantum theory of gauge fields in the framework of perturbation theory was largely completed. In papers by A.A. Slavnov, by J. Taylor, by B. Lee and J. Zinn-Justin, and by G. t'Hooft and M. Veltman, various methods of invariant regularization were developed, the generalized Ward identities were obtained, and a renormalization procedure was constructed in the framework of perturbation theory. This led to the construction of a finite and unitary scattering matrix for the Yang-Mills field.

Since then, the theory of gauge fields has developed rapidly, both theoretically and phenomenologically. Such development led to the construction of a self-consistent theory of weak and electromagnetic interactions based on the Weinberg-Salam model, as well as to a successful description of hadron processes in the region of asymptotic freedom, where one can apply perturbation theory. From a purely theoretical point of view, profound relations were established of gauge theories with differential geometry and topology.

At present the main efforts are directed at the creation of computational methods not related to the expansion in the coupling constant. Along this way promising lines of activity are coming into being that raise great hopes. These hopes, however, have not been fully implemented yet. These include quantization in the neighborhood of nontrivial classical solutions (instantons), computations on large computers in the framework of the lattice approximation, application of methods of the theory of phase transitions, expansion in inverse powers of the number of colors, and a number of other methods.

Approaches are also being developed which combine utilization of the quantum theory of gauge fields and the dispersion technique (sum rules). In brief, hard work aimed at development of the theory of gauge fields is well under way.

From the above short historical survey we shall pass on to the description of the Yang-Mills field itself. For this, we must first recall some notation from the theory of compact Lie groups. More specifically, we shall be interested mainly in the Lie algebras of these groups. Let Ω be a compact semisimple Lie group, that is a compact group which has no invariant commutative (Abelian) subgroups. The number of independent parameters that characterize an arbitrary element of the group (that is, the dimension is equal to n. Among the representations of this group and its Lie algebra, there exists the representation of $n \times n$ matrices (adjoint representation). It is generated by the natural action of the group on itself by the similarity transformations

$$h \to \omega h \omega^{-1}; \quad h, \omega \in \Omega. \tag{1.1}$$

Any matrix $\mathcal{F}$ in the adjoint representation of the Lie algebra can be represented by a linear combination of n generators,

$$\mathcal{F} = T^a \alpha^a. \tag{1.2}$$

For us it is essential that the generators T^a can be normalized by the condition

$$\mathrm{tr}\,(T^a T^b) = -2\delta^{ab}. \tag{1.3}$$

In this case the structure constants t^{abc} that take part in the condition

$$[T^a, T^b] = t^{abc} T^c, \tag{1.4}$$

are completely antisymmetric. The reader unfamiliar with the theory of Lie groups may keep in mind just these two relationships, which are actually a characterizing property of the compact semisimple Lie group.

A compact semisimple group is called simple if it has no invariant Lie subgroups. A general semisimple group is a product of simple groups. This means that the matrices of the Lie algebra in the adjoint representation have a blocked-diagram form, where each block corresponds to one of the simple factors. The generators of the group can be chosen so that each one has nonzero matrix elements only within one of the blocks. We shall always have in mind exactly such a choice of generators, in correspondence with the structure of the direct product.

The simplest example of such a group is the simple group $SU(2)$. The dimension of this group equals 3, and the Lie algebra in the adjoint representation is given by the antisymmetric 3×3 matrices; as generators the matrices

$$T^1 = \begin{pmatrix} 0 & 0 & 0 \\ 0 & 0 & -1 \\ 0 & 1 & 0 \end{pmatrix}; \quad T^2 = \begin{pmatrix} 0 & 0 & 1 \\ 0 & 0 & 0 \\ -1 & 0 & 0 \end{pmatrix}; \quad T^3 = \begin{pmatrix} 0 & -1 & 0 \\ 1 & 0 & 0 \\ 0 & 0 & 0 \end{pmatrix}; \tag{1.5}$$

can be chosen; the structure constants t^{abc} in this base coincide with the completely antisymmetric tensor ε^{abc}.

Besides semisimple compact groups, we shall also deal with the commutative (Abelian) group $U(1)$. The elements of this group are complex numbers, with absolute values equal to unity. The Lie algebra of this group is one-dimensional and consists of imaginary numbers or of real antisymmetric 2×2 matrices.

The Yang-Mills field can be associated with any compact semisimple Lie group. It is given by the vector field $\mathcal{A}_\mu(x)$, with values in the Lie algebra of this group. It is convenient to consider $\mathcal{A}_\mu(x)$ to be a matrix in the adjoint representation of this algebra. In this case it is defined by its coefficients $A_\mu^a(x)$:

$$\mathcal{A}_\mu(x) = A_\mu^a(x) T^a \tag{1.6}$$

with respect to the base of the generators T^a.

In the case of the group $U(1)$ the electromagnetic field $\mathcal{A}_\mu(x) = iA_\mu(x)$ is an analogous object.

We shall now pass on to the definition of the gauge group and its action on Yang-Mills fields. In the case of electrodynamics the gauge transformation is actually the well known gradient transformation

$$\mathcal{A}_\mu(x) \rightarrow \mathcal{A}_\mu(x) + i\partial_\mu \lambda(x). \tag{1.7}$$

Let us recall its origin in the framework of the classical field theory. The electromagnetic field interacts with charged fields, which are described by complex functions $\psi(x)$. In the equations of motion the field $\mathcal{A}_\mu(x)$ always appears in the following combination:

$$\nabla_\mu \psi = (\partial_\mu - \mathcal{A}_\mu)\psi = (\partial_\mu - iA_\mu)\psi. \tag{1.8}$$

The above gradient transformation provides the covariance of this combination with respect to the phase transformation of the fields ψ. If ψ transforms according to the rule

$$\begin{aligned} \psi(x) &\rightarrow e^{i\lambda(x)}\psi(x), \\ \bar{\psi}(x) &\rightarrow e^{-i\lambda(x)}\bar{\psi}(x), \end{aligned} \tag{1.9}$$

then $\nabla_\mu \psi$ transforms in the same way. Indeed,

$$(\partial_\mu - \mathcal{A}_\mu)\psi \rightarrow [\partial_\mu - i\partial_\mu \lambda(x) - \mathcal{A}_\mu(x)]e^{i\lambda(x)}\psi(x) = e^{i\lambda(x)}[\partial_\mu - \mathcal{A}_\mu(x)]\psi(x). \tag{1.10}$$

As a result, the equations of motion are also covariant with respect to the transformations (1.7) and (1.9); if the pair $\psi(x), \mathcal{A}_\mu(x)$ is a solution, then $e^{i\lambda(x)}\psi(x)$, $\mathcal{A}_\mu(x) + i\partial_\mu \lambda(x)$ is also a solution.

In other words, a local change in phase of the field $\psi(x)$, which can be considered to be the coordinate in the charge space, is equivalent to the appearance of an additional electromagnetic field. We see here a complete analogy with the weak equivalence principle in Einstein's theory of gravity, where a change of the coordinate system leads to the appearance of an additional gravitational field.

Extending this analogy further, one may formulate the relativity principle in the charge space. This principle was first introduced by H. Weyl in 1919: The field configurations $\psi(x), \mathcal{A}_\mu(x)$ and $\psi(x)e^{i\lambda(x)}, \mathcal{A}_\mu(x) + i\partial_\mu\lambda(x)$ described the same physical situation. If the construction of theory is based on this principle, then the above-described way of constructing the equations of motion in terms of covariant derivatives is the only possible one.

The generalization of this principle to the case of the more complicated charge space leads to the Yang-Mills theory. Examples of such charge (or internal, as they are often called) spaces are the isotopic space, the unitary-spin space in the theory of hadrons, and so on. In all these examples we deal with fields $\psi(x)$ that acquire values in the charge space, which itself is a representation space for some compact semisimple groups $\Omega(SU(2), SU(3)$, etc.). The equations of motion for the fields $\psi(x)$ contain the covariant derivative

$$\nabla_\mu = \partial_\mu - \Gamma(\mathcal{A}_\mu), \tag{1.11}$$

where $\Gamma(\mathcal{A}_\mu)$ is the representation of the matrix $\mathcal{A}_\mu$ corresponding to the given representation of the group Ω. For example, if $\Omega = SU(2)$ and the charge space corresponds to the two dimensional representation, then the above-mentioned generators T^a are represented by the Pauli matrices

$$\Gamma(T^a) = \frac{1}{2i}\tau^a, \tag{1.12}$$

where

$$\tau^1 = \begin{pmatrix} 0 & 1 \\ 1 & 0 \end{pmatrix}, \quad \tau^2 = \begin{pmatrix} 0 & -i \\ i & 0 \end{pmatrix}, \quad \tau^3 = \begin{pmatrix} 1 & 0 \\ 0 & -1 \end{pmatrix}, \tag{1.13}$$

and in this case

$$\Gamma(\mathcal{A}_\mu) = \frac{1}{2i}A^a_\mu \tau^a. \tag{1.14}$$

The transformation of the fields $\psi(x)$ analogous to the local phase transformation in electrodynamics has the following form:

$$\psi(x) \to \psi^\omega(x) = \Gamma[\omega(x)]\psi(x), \tag{1.15}$$

where $\omega(x)$ is a function of x which has its values in the group Ω. It is convenient to consider $\omega(x)$ to be a matrix in the adjoint representation of the group Ω. The derivative (1.11) will be covariant with respect to this transformation if the field $\mathcal{A}_\mu(x)$ transforms according to the rule

$$\mathcal{A}_\mu(x) \to \mathcal{A}^\omega_\mu(x) = \omega(x)\mathcal{A}_\mu(x)\omega^{-1}(x) + \partial_\mu\omega(x)\omega^{-1}(x). \tag{1.16}$$

It is not difficult to see that this transformation obeys the group law. The set of these transformations composes a group that may formally be denoted by

$$\tilde{\Omega} = \prod_x \Omega. \tag{1.17}$$

This group is called the group of gauge transformations.

Often it is convenient to deal with the infinitesimal form of the gauge transformation. Let the matrices $\omega(x)$ differ infinitesimally from the unit matrix

$$\omega(x) = 1 + \alpha(x) = 1 + \alpha^a(x)T^a, \tag{1.18}$$

where $\alpha(x)$ belongs to the Lie algebra of the group Ω. Then the change of $\mathcal{A}_\mu$ under such a transformation will be

$$\delta \mathcal{A}_\mu = \partial_\mu \alpha - [\mathcal{A}_\mu, \alpha] = \nabla_\mu \alpha, \tag{1.19}$$

or for the components,

$$\delta A^a_\mu = \partial_\mu \alpha^a - t^{abc} A^b_\mu \alpha^c. \tag{1.20}$$

The corresponding transformation for ψ takes the form

$$\delta \psi = \Gamma(\alpha)\psi. \tag{1.21}$$

It is obvious that the group of gradient transformations in electrodynamics is a particular case of the gauge group.

The existence of covariant derivatives makes it possible to dynamically realize the relativity principle in the internal space: The field configurations $\psi(x), \mathcal{A}_\mu(x)$ and $\Gamma[\omega(x)]\psi(x), \mathcal{A}^\omega_\mu(x)$ describe the same physical situation. If we take this principle as a basis for constructing the dynamics, we then automatically come to the Yang-Mills theory.

The relativity principle means that not only one set of fields but also a whole class of gauge-equivalent configurations corresponds to the true physical configuration. To be clearer, this principle implies that in the internal charge space there is no special fixed basis with respect to which the physical fields of matter Ω are represented in terms of components: $\psi = (\psi_1, \ldots, \psi_m)$. Such a basis can be introduced localy at each space-time point; however, there is no physical reason for fixing its position. The local change of basis is interpreted as a change of the gauge field, which plays a role analogous to the role of gravitational or electromagnetic fields.

The relativity principle leads to a significant formal difference in the description of the dynamics of gauge fields in comparison with more customary fields such as, for example, the self-interacting scalar field. In order to work in practice with classes of equivalent configurations, they must somehow be parameterized; that is, in each class unique representatives must be chosen. Usually, this is achieved by imposing a subsidiary condition that eliminates the gauge freedom. This subsidiary condition is called the gauge condition, or simply gauge. The most frequently used gauges are the following conditions:

$$\begin{aligned} &\Phi_L \equiv \partial_\mu \mathcal{A}_\mu = 0 \text{ (Lorentz gauge)},\\ &\Phi_C \equiv \partial_k \mathcal{A}_k = 0 \text{ (Coulomb gauge)},\\ &\Phi_H \equiv \mathcal{A}_0 = 0 \quad \text{(Hamilton gauge)},\\ &\Phi_A \equiv \mathcal{A}_3 = 0 \quad \text{(axial gauge)}. \end{aligned} \tag{1.22}$$

For a general system including both fields $\mathcal{A}_\mu$ and fields ψ, the latter may enter into the gauge condition. Examples of such conditions will be presented in Section 1.3.

In general the gauge condition $\Phi(A, \psi; x)$ is a family of functionals of $\mathcal{A}_\mu$ and ψ, one for each x. For fixed $x\Phi(A, \psi; x)$ is an element of the Lie algebra of the group Ω, so the number of independent gauge conditions coincides with the dimension of the gauge group. In the example (1.22) all the conditions are exactly of such a form. Furthermore, in these examples the gauge conditions are local, that is $\Phi(A, \psi; x)$ depends on the values of A_μ and ψ in the neighborhood of point x.

Let us discuss the requirements to be satisfied by the gauge conditions. The most important one implies that the system of equations

$$\Phi(A^\omega, \psi^\omega; x) = 0 \tag{1.23}$$

has a unique solution $\omega(x)$ for fixed A_μ and ψ. This requirement means that in each set of equivalent fields there actually exists a unique set of fields A_μ, ψ which satisfies the condition (1.23). This set, considered as a representative of the class, characterizes uniquely the true physical configuration. Another requirement that is less fundamental, although important practically, is that Equation (1.23) not be too complicated and should give a sufficiently explicit solution $\omega(x)$, at least in the framework of perturbation theory.

Equation (1.23) is a system of nonlinear equations for $\omega(x)$. For local gauge conditions it is a nonlinear system of partial differential equations. For instance, for the Lorentz gauge this system of equations takes the following form:

$$\begin{aligned} \nabla_\mu L_\mu &= \partial_\mu L_\mu - [A_\mu, L_\mu] = -\partial_\mu A_\mu; \\ L_\mu &= \omega^{-1}\partial_\mu \omega, \end{aligned} \tag{1.24}$$

and for small A_μ and $\alpha(x)$ it is rewritten as

$$\Box\alpha - [A_\mu, \partial_\mu \alpha] + \ldots = -\partial_\mu A_\mu, \tag{1.25}$$

where the dots stand for terms of higher order in α. Equation (1.25) can be uniquely solved with respect to α in the framework perturbation theory if the operator $\Box = \partial_\mu \partial_\mu$ is supplied with suitable boundary conditions. Such boundary conditions arise in the description of the dynamics of Yang-Mills fields and will be discussed in Chapter 3. Nevertheless, beyond the domain of perturbation theory for large fields A_μ, the uniqueness of the solution of Equation (1.24) may fail. Discussion of this possibility is not within the scope of this book.

A necessary condition for the solvability of the equations (1.23) is the non-degeneracy of the corresponding Jacobian. Variation of the gauge condition under an infinitesimal gauge transformation of α defines the linear operator M_Φ that acts on α:

$$M_\Phi \alpha = \int \left[\frac{\delta\Phi(A, \psi; x)}{\delta A_\mu(y)} (\partial_\mu \alpha(y) - [A_\mu(y), \alpha(y)]) + \frac{\delta\Phi(A, \psi; x)}{\delta\psi(y)} \Gamma(\alpha(y))\psi(y) \right] dy, \tag{1.26}$$

This operator plays the role of the Jacobian matrix for the condition (1.23). Non-degeneracy of the operator M_Φ,

$$\det M_\Phi \neq 0 \tag{1.27}$$

is a necessary condition for the existence of a unique solution for the system (1.23).

For local gauge conditions M_Φ is a differential operator obtained while linearizing the system (1.23). For example, in the case of the Lorentz gauge condition, $M_{\Phi_L} = M_L$ has the form

$$M_L\alpha - \Box\alpha - \partial_\mu[\mathcal{A}_\mu, \alpha]. \tag{1.28}$$

This operator is uniquely reversible within the framework of perturbation theory provided boundary conditions are introduced. As was noted above, these conditions will be discussed in Chapter 3.

It seems appropriate here to explain the meaning of the determinant of the differential operator M_L. We shall represent M_L in the form

$$M_L = \Box(1 - \Box^{-1}\partial_\mu[\mathcal{A}_\mu, \cdot]), \tag{1.29}$$

where $\Box^{-1}$ is an integral operator, the Green function, defined uniquely upon introduction of some boundary conditions. Thus, M_L is represented as the product of the operator $\Box$, which is independent of $\mathcal{A}_\mu$, and an operator of the form $\mathrm{I}+K$, where K is the Fredholm operator. As the determinant of the operator M_L we shall take $\det(\mathrm{I}+K)$, which may formally be introduced with the aid of the formula

$$\ln \det(\mathrm{I} + K) = \mathrm{Tr} \ln(\mathrm{I} + K). \tag{1.30}$$

We shall call the condition (1.27) the admisibility condition for the gauge condition; it will be frequently discussed further on.

1.2 Geometrical Interpretation of the Yang-Mills Field

The construction described in the previous section allows an elegant geometrical interpretation when the Yang-Mills fields play the same role as the Christoffel symbols in gravitation gheory. Analogously to the latter, the Yang-Mills fields describe parallel translation in the charge space and determine the curvature of this space, the fields $\psi(x)$ being analogs of tensor fields.

A natural geometrical language for the description of this analogy is yielded by the fiber-bundle theory. In this theory the concept of connection in the principle bundle corresponds to the Yang-Mills field. Although the fiber-bundle theory produces the most adequate language for axiomatization of the classical field theory, in this book, which is addressed mainly to physicists, we shall not use it. We just point out that the general concept of connection, equivalent to the Yang-Mills field, appeared in the mathematical literature only in 1950, that is, practically simultaneously with the work of Yang and Mills.

Let us now explain in what sense the Yang-Mills fields determine parallel translation. Let $\gamma(s)$ be a contour in space-time defined by the equation

$$x_\mu = x_\mu(s). \tag{2.1}$$

The vector field $\dot{\gamma}(s)$ with components

$$X_\mu = \frac{dx_\mu}{ds} \tag{2.2}$$

is tangential to the curve $\gamma(s)$ at each of its points. We shall say that the field $\psi(x)$ undergoes parallel translations along $\gamma(s)$ if at each point of the contour,

$$\nabla_\mu \psi(x) \mid_{x=x(s)} X_\mu = 0, \tag{2.3}$$

that is, the covariant derivative in the tangential direction is equal to zero.

Generally speaking, parallel translation along a closed contour changes the field $\psi(x)$. Let us calculate this change for an infinitesimal contour. We shall consider a contour that has the form of a parallelogram with vertices

$$(x, x + \Delta_1 x, x + \Delta_1 x + \Delta_2 x, x + \Delta_2 x).$$

It may be readily verified that if the covariant derivative along this contour is equal to zero, then the total change in $\psi(x)$ corresponding to a whole turn round the closed contour is equal to

$$\Delta_{12}\psi(x) = \Gamma(\mathcal{F}_{\mu\nu})\psi(\Delta_1 x_\mu \Delta_2 x_\nu - \Delta_1 x_\nu \Delta_2 x_\mu), \tag{2.4}$$

where

$$\mathcal{F}_{\mu\nu} = \partial_\nu \mathcal{A}_\mu - \partial_\mu \mathcal{A}_\nu + [\mathcal{A}_\mu, \mathcal{A}_\nu]. \tag{2.5}$$

Indeed, since the covariant derivative along the side $(x, x + \Delta_1 x)$ is equal to zero, the change in $\psi(x)$ corresponding to the change of x along the first contour equals

$$\Delta_1\psi(x) = \psi(x + \Delta_1 x) - \psi(x) = \partial_\mu \psi \Delta_1 x_\mu = \Gamma(\mathcal{A}_\mu)\psi(x)\Delta_1 x_\mu. \tag{2.6}$$

Performing analogous calculations for the remaining sides of the parallelogram, and taking into account the linear dependence of $\Gamma(\mathcal{A}_\mu)$ upon $\mathcal{A}_\mu$ and the fact that $[\Gamma(\mathcal{A}_\mu), \Gamma(\mathcal{A}_\nu)] = \Gamma([\mathcal{A}_\mu, \mathcal{A}_\nu])$ we obtain the formula (2.4) for the total change in $\psi(x)$. This formula shows that it is natural to call $\mathcal{F}_{\mu\nu}$ the curvature of the charge space.

Under gauge transformations, $\Delta\psi(x)$ changes in the same way as $\psi(x)$. This is because for the construction of $\Delta\psi(x)$ we have used only the covariant derivative. Then from (2.4) it follows that $\Gamma(\mathcal{F}_{\mu\nu}(x))$ transforms according to the law

$$\Gamma(\mathcal{F}_{\mu\nu}(x)) \to \Gamma(\omega(x))\Gamma(\mathcal{F}_{\mu\nu}(x))\Gamma(\omega^{-1}(x)). \tag{2.7}$$

Therefore $\mathcal{F}_{\mu\nu}(x)$ itself under gauge transformations transforms as

$$\mathcal{F}_{\mu\nu}(x) \to \omega(x)\mathcal{F}_{\mu\nu}(x)\omega^{-1}(x). \tag{2.8}$$

If we adopt the convention that $\psi(x)$ is a vector with respect to gauge transformations, the $\Gamma(\mathcal{F}_{\mu\nu}(x))$ is a tensor of rank two. And $\mathcal{F}_{\mu\nu}(x)$ itself is sometimes conveniently considered a vector in the adjoint representation.

Our indirect derivation of (2.8) is verified by a straightforward check if one takes advantage of the explicit expression (2.5) for $\mathcal{F}_{\mu\nu}(x)$ in terms of $\mathcal{A}_\mu(x)$ and of the transformation law (1.16) for $\mathcal{A}_\mu(x)$. Thus we conclude our short description of the geometrical interpretation of the Yang-Mills fields: They describe parallel translation of vectors in the charge space, and the tensor $\mathcal{F}_{\mu\nu}(x)$ is the curvature tensor of this space. The reader familiar with the theory of gravity must surely have already noticed the complete analogy between $\mathcal{A}_\mu(x)$ and Christoffel symbols, and between $\mathcal{F}_{\mu\nu}(x)$ and the curvature tensor of the gravitational field. To conclude this analogy, we point out that the tensor $\mathcal{F}_{\mu\nu}(x)$ is the commutator of the covariant derivatives

$$\mathcal{F}_{\mu\nu}(x) = [\nabla_\mu, \nabla_\nu] \tag{2.9}$$

and the Jacobi identity

$$[[\nabla_\mu, \nabla_\nu], \nabla_\sigma] + \text{cyclic permutations} = 0 \tag{2.10}$$

leads to the identity

$$\nabla_\sigma \mathcal{F}_{\mu\nu}(x) + \text{cyclic permutations} = 0, \tag{2.11}$$

where $\nabla_\sigma \mathcal{F}_{\mu\nu}(x) = \partial_\sigma \mathcal{F}_{\mu\nu}(x) - [\mathcal{A}\sigma(x), \mathcal{F}_{\mu\nu}(x)]$ which is the analog of the Bianchi identity in the theory of gravity. A similar consideration can be carried out in the case of the Abelian group U(1). In this case

$$\mathcal{F}_{\mu\nu}(x) = \partial_\nu \mathcal{A}_\mu(x) - \partial_\mu \mathcal{A}_\nu(x) = i(\partial_\nu A_\mu(x) - \partial_\mu A_\nu(x)), \tag{2.12}$$

which obviously coincides with the tensor of the elecltromagnetic field strength. The interpretation of $\mathcal{F}_{\mu\nu}(x)$ as the curvature of the charge space, originating with V. Fock and H. Weyl, is the most natural approach to the geometrization of the electromagnetic field. Numerous attempts to relate this field to the geometrical properties of space-time itself have never had any success.

In concluding this section we shall say a few words about the classical dynamics of the Yang-Mills field. Our task is to construct a gauge-invariant Lagrange function coinciding, in the case of the Abelian group $U(1)$, with the Lagrangian of the electromagnetic field

$$\mathcal{L} = \frac{1}{4e^2} \mathcal{F}_{\mu\nu} \mathcal{F}_{\mu\nu} + \mathcal{L}_M(\psi, \nabla_\mu \psi), \tag{2.13}$$

where $\mathcal{L}_M$ describes the gauge-invariant interaction of the fields $\mathcal{A}_\mu(x)$ and $\psi(x)$ and is deduced from the free Lagrangian of the fields ψ by replacing ordinary derivatives with covariant ones, and where e plays the role of the electric charge. This formula may be easily rewritten in a more familiar form if one changes the normalization of the fields:

$$\mathcal{A}_\mu(x) \to e\mathcal{A}_\mu(x). \tag{2.14}$$

In this case the factor e^{-2} vanishes from the first term but appears instead in the expression for the covariant derivative,

$$\nabla_\mu \to \partial_\mu - e\mathcal{A}_\mu.$$

In the following we shall use both methods of normalizing the fields $\mathcal{A}_\mu(x)$ without specially mentioning this.

A natural (and the only possible) generalization of the formula (2.13) to the case of the simple non-Abelian gauge group is the following expression:

$$\mathcal{L} = \frac{1}{8g^2} \operatorname{tr} \mathcal{F}_{\mu\nu}\mathcal{F}_{\mu\nu} + \mathcal{L}_M(\psi, \nabla_\mu \psi). \tag{2.15}$$

The first term may be rewritten also as

$$\mathcal{L} = -\frac{1}{4g^2} F^a_{\mu\nu} F^a_{\mu\nu}, \tag{2.16}$$

where $\mathcal{F}^a_{\mu\nu}(x)$ are the components of the matrix $\mathcal{F}_{\mu\nu}(x)$ with respect to the base T^a. Obviously, this Lagrangian is invariant with respect to the gauge transformations (1.15), and (1.16). In the case of the semisimple group of general form, the Lagrangian contains r arbitrary constants $g_i, i = 1, \ldots, r$, where r is the number of invariant simple factors. Then the formula analogous to (2.16) takes the form

$$\mathcal{L} = \sum_i -\frac{1}{4g_i^2} F^{a_i}_{\mu\nu} F^{a_i}_{\mu\nu}, \tag{2.17}$$

where i is the index number of a simple factor.

Contrary to electrodynamics, the Lagrangian (2.16) of the Yang-Mills field in vacuum (that is, in the absence of the fields ψ), in addition to the second-order terms in the fields, contains higher-order terms. This means that Yang-Mills fields have nontrivial self-interaction. In other words, quanta of the Yang-Mills field themselves have charges, the interaction of which they transfer. The main specific feature of the Yang-Mills field dynamics is related to this self-action; therefore we shall often confine ourselves to the model of the Yang-Mills field in vacuum when dealing with general problems.

The equations of motion arising from the Lagrangian (2.16) for the Yang-Mills field in vacuum have the form

$$\nabla_\mu \mathcal{F}_{\mu\nu} = \partial_\mu \mathcal{F}_{\mu\nu} - [\mathcal{A}_\mu, \mathcal{F}_{\mu\nu}] = 0 \tag{2.18}$$

and written in terms of the $\mathcal{A}_\mu$,

$$\Box \mathcal{A}_\nu - \partial_\nu \partial_\mu \mathcal{A}_\mu + [\mathcal{A}_\mu, (\partial_\nu \mathcal{A}_\mu - \partial_\mu \mathcal{A}_\nu + [\mathcal{A}_\mu, \mathcal{A}_\nu])] - \partial_\mu [\mathcal{A}_\mu, \mathcal{A}_\nu] = 0, \tag{2.19}$$

represent a system of second-order equations. These equations are gauge-invariant in the following sense: If $\mathcal{A}_\mu$ is a solution of (2.19), then $\mathcal{A}^\omega_\mu$ is also a solution for any arbitrary $\omega(x)$. This means that the standard parameterization of the solutions in terms of the initial conditions $(\mathcal{A}_\mu(x,t), \partial_o \mathcal{A}_\mu(x,t)$ at a fixed $t)$ is unfit for the system (2.19). If we impose the gauge conditions, this obstacle is eliminated; however, the initial conditions are then not arbitrary but are restricted by the gauge conditions.

Models of interaction of the Yang-Mills field with fields of matter will be considered in the next section.

1.3 Dynamical Models with Gauge Fields

The Lagrangian describing the interaction of the Yang-Mills field with spinor fields is the simplest one. Let a multiplet of spinor fields $\psi_k(x)$ realize the representation $\Gamma(\omega)$ of a simple, compact gauge group Ω. Then the Lagrangian has the form

$$\mathcal{L} = \mathcal{L}_{YM} + i\bar{\psi}(x)\gamma_\mu \nabla_\mu \psi(x) - m\bar{\psi}(x)\psi(x). \tag{3.1}$$

We have used here the following notation: $\mathcal{L}_{YM}$ is the already familiar Lagrangian of the Yang-Mills field in vacuum,

$$\mathcal{L}_{YM} = \frac{1}{8g^2} \operatorname{tr} \mathcal{F}_{\mu\nu}\mathcal{F}_{\mu\nu}. \tag{3.2}$$

In the scalar product of two spinors the sum is over the indices corresponding to internal degrees of freedom; for instance, the mass term may be written down as

$$m\bar{\psi}(x)\psi(x) = m\bar{\psi}_k(x)\psi_k(x). \tag{3.3}$$

Furthermore,

$$(\nabla_\mu \psi(x))_k = \partial_\mu \psi_k(x) - (\Gamma(\mathcal{A}_\mu(x)))_{kl}\psi_l(x), \tag{3.4}$$

where $(\Gamma(\mathcal{A}_\mu))_{kl} = A^a_\mu(\Gamma(T^a))_{kl}$ and the matrix $(\Gamma(T^a))_{kl}$ which in the following will be denoted simply by Γ^a_{kl} is the matrix of the generator T^a in the representation realized by the fields $\psi(x)$. Then

$$\bar{\psi}(x)\gamma_\mu \nabla_\mu \psi(x) = \bar{\psi}_k(x)\gamma_\mu(\partial_\mu \psi_k(x) - A^a_\mu(x)\Gamma^a_{kl}\psi_l(x)). \tag{3.5}$$

For example, let the gauge group be $\Omega = SU(2)$, and let the fields $\psi(x)$ realize the fundamental representation of this group. Then

$$(\Gamma(\mathcal{A}_\mu))_{kl} = -\frac{i}{2}A^a_\mu(\tau^a)_{kl}, \tag{3.6}$$

where τ^a are the Pauli matrices, and the complete Lagrangian has the form

$$\mathcal{L} = -\frac{1}{4g^2}(\partial_\nu A^a_\mu - \partial_\mu A^a_\nu + \varepsilon^{abc}A^b_\mu A^c_\nu)^2 + i\bar{\psi}\gamma_\mu\left(\partial_\mu\psi + \frac{i}{2}A^a_\mu\tau^a\psi\right) - m\bar{\psi}\psi. \tag{3.7}$$

In the case when the gauge group is the group $SU(3)$ and the spinors $\psi(x)$ realize its fundamental (spinor) representation, the analogous Lagrangian takes the form

$$\mathcal{L} = -\frac{1}{4g^2}(\partial_\nu A^x_\mu - \partial_\mu A^a_\nu + f^{abc}A^b_\mu A^c_\nu)^2 + i\bar{\psi}\gamma_\mu\left(\partial_\mu\psi + \frac{i}{2}A^a_\mu\gamma^a\psi\right) - m\bar{\psi}\psi, \tag{3.8}$$

where f^{abc} are the structure constants of the group $SU(3)$ and the matrices λ^a are the well-known Gell-Mann matrices:

$$\lambda_1 = \begin{pmatrix} 0 & 1 & 0 \\ 1 & 0 & 0 \\ 0 & 0 & 0 \end{pmatrix}; \quad \lambda_2 = \begin{pmatrix} 0 & -i & 0 \\ i & 0 & 0 \\ 0 & 0 & 0 \end{pmatrix}; \quad \lambda_3 = \begin{pmatrix} 1 & 0 & 0 \\ 0 & -1 & 0 \\ 0 & 0 & 0 \end{pmatrix};$$
$$\lambda_4 = \begin{pmatrix} 0 & 0 & 1 \\ 0 & 0 & 0 \\ 1 & 0 & 0 \end{pmatrix}; \quad \lambda_5 = \begin{pmatrix} 0 & 0 & -i \\ 0 & 0 & 0 \\ i & 0 & 0 \end{pmatrix}; \quad \lambda_6 = \begin{pmatrix} 0 & 0 & 0 \\ 0 & 0 & 1 \\ 0 & 1 & 0 \end{pmatrix}; \tag{3.9}$$
$$\lambda_7 = \begin{pmatrix} 0 & 0 & 0 \\ 0 & 0 & -i \\ 0 & i & 0 \end{pmatrix}; \quad \lambda_8 = \frac{1}{\sqrt{3}} \begin{pmatrix} 1 & 0 & 0 \\ 0 & 1 & 0 \\ 0 & 0 & -2 \end{pmatrix}.$$

Renormalization of the fields

$$A_\mu^a(x) \to g A_\mu^a(x) \tag{3.10}$$

changes the form of Lagrangians (3.7) and (3.8) to a more familiar one, where g is involved only in the interaction term.

The latter Lagrangian is used, for example, in the theory of strong interactions. In this case the spinors ψ are identified with the quark fields, the Yang-Mills fields are called gluons, and the internal space is called the space of colors.

In terms of the variables (3.10) the equations of motion in quantum chromodynamics have the form

$$\partial_\mu F_{\mu\nu}^a - g[A_\mu, \mathcal{F}_{\mu\nu}]^a = g\bar{\psi}\gamma^\mu \frac{\lambda^a}{2}\psi, \tag{3.11}$$

$$i\gamma_\mu \left(\partial_\mu \psi + i\frac{g}{2} A_\mu^a \lambda^a \psi\right) - m\psi = 0, \tag{3.12}$$

$$-i\left(\partial_\mu \bar{\psi} - i\frac{g}{2}\bar{\psi} A_\mu^a \lambda^a\right)\gamma_\mu - m\psi = 0. \tag{3.13}$$

In the above examples, when the gauge group is simple, all interactions are characterized by a single coupling constant. Such universality of the interactions is a specific feature of the Yang-Mills theory.

The next useful example is the interaction of the Yang-Mills field with a scalar field. Let the multiplet of scalar fields $\varphi_k(x)$ realize a real representation $\Gamma(\omega)$ of the simple compact group Ω. Then the gauge-invariant Lagrangian has the form

$$\mathcal{L} = \mathcal{L}_{YM} + \frac{1}{2}\nabla_\mu\varphi\nabla_\mu\varphi - \frac{m^2}{2}\varphi\varphi - V(\varphi), \tag{3.14}$$

where the covariant derivative $\nabla_\mu\varphi$ is constructed as above

$$\nabla_\mu\varphi = \partial_\mu\varphi - \Gamma(A_\mu)\varphi, \tag{3.15}$$

where $\varphi\varphi$, as before, is a scalar product in the charge space, and $V(\varphi)$ is a form of third and fourth degree in the fields φ that is invariant with respect to the group Ω.

In the case when $\Omega = SU(2)$ and fields φ realize the adjoint representation $\varphi = \varphi^a, a = 1,2,3$, the corresponding formula becomes

$$\mathcal{L} = \mathcal{L}_{YM} + \frac{1}{2}(\partial_\mu \varphi^a - g\varepsilon^{abc} A^b_\mu \varphi^c)^2 - \frac{m^2}{2}\varphi^a\varphi^a - \lambda^2(\varphi^a\varphi^a)^2, \tag{3.16}$$

where parameters m and λ^2 play the role of the masses and of the contact-interaction coupling constants of the scalar fields. The Lagrangian (3.13) itself is, evidently, of little interest from the viewpoint of physical applications; however, an insignificant-looking modification leads to the extremely interesting possibility of describing massive vector fields within the framework of the Yang-Mills theory. This mechanism for the mass generation of the vector field is called the *Higgs effect*. We now proceed to discuss this effect.

We shall continue to deal with the gauge group $SU(2)$ as an example. We first cnsider the case when the scalar field belongs to the adjoint representation. We shall use the following Lagrangian:

$$\mathcal{L} - \mathcal{L}_{YM} + \frac{1}{2}(\nabla_\mu \varphi^a)^2 - \lambda^2(\varphi^a\varphi^a - \mu^2)^2. \tag{3.17}$$

This Lagrangian differs from the Lagrangian (3.13), which we examined earlier by the constant term $-\lambda^2\mu^4$ and the sign of the term with φ squared. At first sight, the Lagrangian (3.14) appears to describe particles with imaginary masses and therefore seems to have no physical meaning. Such a conclusion, however, would be too hasty. The term with φ^2 plays the role of the mass only if $\varphi = 0$ is the stable equilibrium point and is, therefore, the potential-energy minimum. In our case the potential energy is

$$U(A_\mu, \varphi) = \int \Big[\frac{1}{4g^2}F^a_{ik}F^a_{ik} + \frac{1}{2}\nabla_i\varphi^a\nabla_i\varphi^a$$
$$+\lambda^2(\varphi^a\varphi^a - \mu^2)^2\Big]\, d^3x, \quad i,k = 1,2,3, \tag{3.18}$$

and the configuration $\varphi^a = 0$, $A^a_\mu = 0$ is a saddle point. The corresponding equilibrium is unstable. However, stable equilibrium points also exist; they are the configurations corresponding to zero A^a_μ and to constant φ with a fixed length $\varphi^2 = \mu^2$. Such A_μ, φ nullify all the three positive terms that make up the potential energy. (It should be pointed out that besides these configurations themselves, gauge transformation of these configurations obviously yields configurations that are also minima. However, owing to the relativity principle, these configurations present no new physical information, and we shall not take them into consideration.)

Besides these translation-invariant minima the potential energy has other ones, for example, minima corresponding to the 't Hooft-Polyakov monopoles. The energy values for these configurations, however, are higher, so they are only local minima.

In order to determine the real masses, it is necessary to expand the potential energy in Taylor's series around the true minimum. In our case the equilibrium point is degenerate. The minimal configurations form a two-dimensional sphere S^2 with points corresponding to directions of the constant vector φ. We shall note these directions by $\vec{n}$ and write the corresponding φ with the index n, so that $\vec{\varphi}_n = \mu\vec{n}$. The degeneracy is eliminated if we reduce the configuration space and take into consideration only fields φ which coincide asymptotically with one of the φ_n at high $|x|$. Such a choice, naturally, destroys the invariance under $SU(2)$ transformations with constant parameters (isotopic invariance). It may be shown that this condition does not contradict the dynamics and that theories corresponding to different choices of φ_n are physically equivalent. The reader familiar with solid-state physics will, of course, see here an analogy to the theory of ferromagnetics, in which a choice of the direction of the spontaneous magnetization must be made in order to formulate the theory itself.

For definiteness, let us choose the vector $\vec{n}$ to be directed along the third axis: $\vec{n} = (0, 0, 1)$. The corresponding vector $\vec{\varphi}_n$ is $(0, 0, \mu)$.

Transition of fields $\varphi(x)$ with a zero asymptotics at infinity,

$$\varphi(x) \rightarrow \varphi_n + \varphi(x) \tag{3.19}$$

makes the isotopic-symmetry breaking explicit, and the Lagrangian takes the form

$$\begin{aligned}\mathcal{L} = \mathcal{L}_{YM} &+ \frac{1}{2}(\nabla_\mu\varphi^a)^2 + \frac{m_1^2}{2}[(A_\mu^1)^2 + (A_\mu^2)^2] + m_1(A_\mu^1\partial_\mu\varphi^2 - A_\mu^2\partial_\mu\varphi^1)\\ &+ gm_1[\varphi^3[(A_\mu^1)^2 + (A_\mu^2)]^2 - A_\mu^3[\varphi^1 A_\mu^1 + \varphi^2 A_\mu^2]]\\ &- \frac{m_2^2}{2}(\varphi^3)^2 - \frac{m_2^2 g}{2m_1}\varphi^3(\varphi^a)^2 - \frac{m_2^2 g^2}{8m_1^2}(\varphi^a\varphi^a)^2,\\ &m_1 = \mu g; \quad m_2 = 2\sqrt{2}\lambda\mu.\end{aligned} \tag{3.20}$$

Although we have explicity broken isotopic invariance, the Lagrangian and the boundary conditions are invariant under local gauge transformations with functions $\omega(x)$ tending to unity at infinity. We shall give the explicit form of the gauge transformations in new variables, confining ourselves to infinitesimal transformations:

$$\delta\varphi^a(x) = -g\varepsilon^{abc}\varphi^b(x)\alpha^c(x) - m_1\varepsilon^{a3c}\alpha^c(x). \tag{3.21}$$

In order to analyze the spectrum of masses generated by the Lagrangian (3.20), one must choose representatives in gauge-equivalent classes of fields $A_\mu(x), \varphi(x)$ that is, one must fix the gauge. It is convenient to choose the folowing gauge condition:

$$\varphi^1(x) = 0; \quad \varphi^2(x) = 0; \quad \partial_\mu A_\mu^3(x) = 0. \tag{3.22}$$

It can be verified that for sufficiently small $\varphi^3(x)$ the admissibility condition is fulfilled. Indeed,

$$\delta(\partial_\mu A_\mu^3) = \Box\alpha^3 - g\varepsilon^{3bc}\partial_\mu[A_\mu^b\alpha^c] \tag{3.23}$$

and $\delta\varphi^{1,2}$ are determined by the formula (3.18). As a result, the operator M corresponding to our gauge has the form

$$M\begin{pmatrix}\alpha^1\\ \alpha^2\\ \alpha^3\end{pmatrix}=\begin{pmatrix}0, & -g\varphi^3-m_1, & g\varphi^2\\ -g\varphi^3-m_1 & 0, & -g\varphi^1\\ \partial_\mu A_\mu^2+A_\mu^2\partial_\mu, & -A_\mu^1\partial_\mu-\partial_\mu A_\mu^1, & \Box\end{pmatrix}\begin{pmatrix}\alpha_1\\ \alpha_2\\ \alpha_3\end{pmatrix}. \quad (3.24)$$

At small φ the determinant of the operator M is

$$\det M = m_1^2 \det\Box + O(\varphi). \quad (3.25)$$

Since the first term is not zero, in the framework of perturbation theory $\det M \neq 0$, and the admissibility condition is fulfilled.

We shall now write explicitly the quadratic form determining the mass spectrum:

$$\mathcal{L}_0=-\frac{1}{4}(\partial_\nu A_\mu^a-\partial_\mu A_\nu^a)^2+\frac{m_1^2}{2}((A_\mu^1)^2+(A_\mu^2)^2+\frac{1}{2}\partial_\mu\varphi^3\partial_\mu\varphi^3-\frac{m_2^2}{2}(\varphi^3)^2. \quad (3.26)$$

As is seen, our theory in the classical approximation describes two massive vector fields, one massless vector field, and one massive scalar particle. Therefore, indeed, two vector fields have acquired masses; however, quanta of two scalar fields have disappeared from the list of particles.

It is not difficult to construct an $SU(2)$-gauge-invariant model in which all three vector fields acquire a nonzero mass. For this it is necessary to examine the complex scalar field multiplet in the two-dimensional (spinor) representation.

$$\varphi=\begin{pmatrix}\varphi_1\\ \varphi_2\end{pmatrix}, \quad \varphi^+=(\varphi_1^*,\varphi_2^*). \quad (3.27)$$

The gauge-invariant Lagrangian has the form

$$\mathcal{L}=\mathcal{L}_{YM}+(\nabla_\mu\varphi)^+\nabla_\mu\varphi-\lambda^2(\varphi^+\varphi-\mu^2)^2, \quad (3.28)$$

where

$$\nabla_\mu\varphi=\partial_\mu\varphi+\frac{i}{2}g\tau^a A_\mu^a\varphi, \quad (3.29)$$

and the gauge transformation of the fields φ is given by the formula

$$\delta\varphi(x)=\frac{1}{2i}g\tau^a\alpha^a(x)\varphi(x). \quad (3.30)$$

As in the previous case, a stable extremum corresponds to a constant φ such that

$$\varphi^+\varphi=\mu^2. \quad (3.31)$$

We see that in this case the set of stable extrema forms a three-dimensional sphere S^3. In order to remove the degeneracy, we choose as a minimum

$$\varphi(x)=\begin{pmatrix}0\\ \mu\end{pmatrix}. \quad (3.32)$$

It can be verified that the condition

$$\varphi_1(x)=0; \qquad \operatorname{Im}\varphi_2(x)=0 \quad (3.33)$$

is an admissible gauge. In this gauge there remains only one scalar field

$$\operatorname{Re}\varphi_2(x) = \frac{1}{\sqrt{2}}\sigma(x).$$

Passing to fields with zero asymptotics at infinity,

$$\sigma(x) \to \sqrt{2}\mu + \sigma(x), \tag{3.34}$$

we obtain the Lagrangian

$$\mathcal{L} = -\frac{1}{4}F^a_{\mu\nu}F^a_{\mu\nu} + \frac{m_1^2}{2}A^a_\mu A^a_\mu + \frac{1}{2}\partial_\mu\sigma\partial_\mu\sigma - \frac{1}{2}m_2^2\sigma^2$$
$$+ \frac{m_1 g}{2}\sigma A^a_\mu A^a_\mu + \frac{g^2}{8}\sigma^2 A^a_\mu A^a_\mu - \frac{gm_2^2}{4m_1}\sigma^3 - \frac{g^2 m_2^2}{32m_1^2}\sigma^4,$$
$$m_1 = \frac{\mu g}{\sqrt{2}}; \quad m_2 = 2\lambda\mu, \tag{3.35}$$

which describes the interaction of three massive vector fields and one massive scalar field.

The above mechanism will further be used for construction of unified gauge-invariant models of weak and electromagnetic interactions. We have finished the discussion of the classical Yang-Mills theory, and we shall now proceed to its quantization.

2

Quantum Theory in Terms of Path Integrals

Several approaches to the quantization of a field theory exist. Most frequently, quantization is carried out by the operator method, in which operators satisfying canonical commutation relations correspond to classical field configurations. There is, however, another approach, in which the quantum dynamics is described by the sum over all field configurations, known as the path integral. Within this approach, R. Feynman first formulated a self-consistent, manifestly relativistic-invariant perturbation theory for quantum electrodynamics. This formalism has turned out to be most convenient for the quantization of gauge fields, since the relativity principle is taken into account in the simplest way: Integration must be performed not over all field configurations but only over gauge-equivalent classes.

In this chapter we deal with the general formalism of the path integral. In the next chapter we shall discuss applications of this formalism to gauge fields.

2.1 The Path Integral over Phase Space

We shall start by demonstrating the main ideas of the path-integral method, as applied, for example, to nonrelativistic quantum mechanics. We begin with the case of a system with one degree of freedom.

Let p and q be the canonical momentum and coordinate of a particle ($-\infty < p < \infty, -\infty < q < \infty$). In the operator method of quantization, corresponding to p and q there are operators P, Q for which, most frequently, two realization are used - the coordinate and the momentum ones. In the coordinate representation these operators and their eigenfunctions have the form

$$Q = x; \qquad P = \frac{1}{i}\frac{d}{dx};$$
$$|q\rangle = \delta(x-q); \qquad |p\rangle = \left(\frac{1}{2\pi}\right)^{1/2} e^{ipx}; \tag{1.1}$$
$$Q|q\rangle = q|q\rangle; \qquad P|p\rangle = p|p\rangle.$$

The transformation functions from the coordinate to the momentum representation and vice versa are given by the formulas

$$\langle p|q\rangle = \left(\frac{1}{2\pi}\right)^{1/2} e^{-ipq}; \qquad \langle q|p\rangle = \left(\frac{1}{2\pi}\right)^{1/2} e^{ipq}. \tag{1.2}$$

The dynamics of the system is described with the help of the Hamiltonian function $h(p,q)$. In quantum mechanics there is a correspondence between this function and the Hamiltonian operator

$$H = h(P,Q), \tag{1.3}$$

Here we assume a certain procedure for ordering noncommuting operator arguments P and Q. We do not discuss here the general problem of ordering, but we shall come back to it when we introduce the concept of the path integral. The formal reasoning we use does not depend upon the concrete choice of the ordering procedure.

For definiteness, we shall apply the symmetric method of ordering, first introduced by H. Weyl.

In the Weyl quantization an operator $F = f(P,Q)$ is made to correspond to the classical observable $f(p,q)$ in accordance with the following rule. Consider the two-parameter set of unitary operators

$$u(s,t) = e^{iPs+iQt} \tag{1.4}$$

and introduce the Fourier transform of the function $f(p,q)$:

$$f(p,q) = \frac{1}{2\pi}\int \tilde{f}(s,t)e^{isp+itq}ds\,dt. \tag{1.5}$$

Then

$$F = \frac{1}{2\pi}\int \tilde{f}(s,t)u(s,t)ds\,dt. \tag{1.6}$$

In the coordinate representation the kernel $\langle q''|F|q'\rangle$ of the operator $f(P,Q)$ is given by the formula

$$\langle q''|F|q'\rangle = \frac{1}{2\pi}\int f\left(p, \frac{q''+q'}{2}\right) e^{ip(q''-q')}dp. \tag{1.7}$$

To obtain this representation, it suffices to note that the kernel of the operator $u(s,t)$ in the coordinate representation has the form

$$\langle q''|u(s,t)|q'\rangle = \exp\left\{i\frac{q''+q'}{2}t\right\}\delta(q''-q'+s). \tag{1.8}$$

Equality (1.8) is readily derived with the aid of the Hausdorf formula

$$e^{A+B} = e^{A}e^{B}e^{\frac{1}{2}[A,B]}, \tag{1.9}$$

valid for arbitrary operators A and B, the commutator of which is proportional to the unit operator.

The evolution of a system in time is determined by the evolution operator

$$U(t'',t') = \exp\{-iH(t''-t')\}. \tag{1.10}$$

For example, within the Schrödinger picture, if at a moment t' a system were to be found in the state ψ', then at the moment t'' it will turn out to be in the state

$$\Psi'' = U(t'',t')\Psi'. \tag{1.11}$$

Now, we shall compute the matrix element of the evolution operator

$$\langle q''|U(t'',t')|q'\rangle = \langle q''|\exp\{-iH(t''-t')\}|q'\rangle, \tag{1.12}$$

which can be called the kernel of the operator U in the coordinate representation. Sometimes this matrix element is termed the transition function and is denoted as $\langle q'',t''|q',t'\rangle$.

For small $t''-t'$ we have

$$\exp\{-iH(t''-t')\} \cong 1 - iH(t''-t'), \tag{1.13}$$

and from formula (1.7) we obtain the kernel of the transition operator in the form

$$\begin{aligned}\langle q''|e^{iH(t''-t')}|q'\rangle &\cong \frac{1}{2\pi}\int e^{ip(q''-q')}\left(1 - ih\left(p,\frac{q''+q'}{2}\right)\times(t''-t')\right)\\ &\cong \frac{1}{2\pi}\int \exp\left\{ip(q''-q') - ih\left(p,\frac{q''+q'}{2}\right)(t''-t')\right\}dp.\end{aligned} \tag{1.14}$$

For a finite interval $t''-t'$ this formula is, of course, incorrect. In this case one may proceed as follows. We divide the interval $t''-t'$ into N steps, considering

$$\Delta t = \frac{t''-t'}{N} \tag{1.15}$$

to be sufficiently small to use the previous formula for the operator $\exp\{-iH\Delta t\}$. The operator $U(t'',t')$ is expressed in terms of $\exp\{-iH\Delta t\}$ by the formula

$$U(t'',t') = (\exp\{-iH\Delta t\})^{n}. \tag{1.16}$$

Replacing each factor to the right by its kernel and integrating over all intermediate states, we obtain

$$\langle q''|U(t'',t')|q'\rangle = \int \exp\Big\{ i[p_N(q_N - q_{N-1}) + \cdots + p_1(q_1 - q_0)] - \left[h\left(p_n, \frac{q_N + q_{N-1}}{2}\right) + \cdots + h\left(p_1, \frac{q_1 + q_0}{2}\right)\right] \Delta t \Big\} \frac{dp_n}{2\pi} \frac{dp_{N-1}dq_{N-1}}{2\pi} \cdots \frac{dp_1 dq_1}{2\pi}, \tag{1.17}$$

where $q_N = q'', q_0 = q'$.

Now we pass over the limit $N \to \infty$, $\Delta t \to 0$. The number of integration variables also approaches infinity, and we can consider that in the limit we integrate over the values of the function $p(t), q(t)$ for all t in the interval $t' < t < t''$. The function $q(t)$ is subject to the condition

$$q(t') = q'; \qquad q(t'') = q''. \tag{1.18}$$

The argument of the exponential function in this limit transforms into the integral

$$A_{t'}^{t''} = \int_{t'}^{t''} (p(t)\dot{q}(t) - h(p(t), q(t)))dt, \tag{1.19}$$

that is, into the classical action on the interval (t', t''). Thus, we obtain the main result: The matrix element of the evolution operator is found by integrating the Feynman functional $\exp\{iA_{t'}^{t''}\}$ over all trajectories $p(t)$, $q(t)$ in the phase space with fixed values q' and q'' at $t = t'$ and $t = t''$, respectively. The integration measure may be written formally as

$$\frac{dp''}{2\pi} \prod_t \frac{dp(t)dq(t)}{2\pi}. \tag{1.20}$$

Thus, if one-dimensional integration over the final momentum $p(t'')$ is dropped, then the quantum-mechanical transition amplitude $< q'', t''|q', t' >$ is written down completely in terms of canonical invariants of classical mechanics, namely, in terms of the action functional and Liouville measure.

The same final result

$$\langle q'', t''|q', t'\rangle = \langle q''|U(t'', t')|q'\rangle = \int \exp\left\{ i \int_{t'}^{t''} (p\dot{q} - h(p, q))dt \right\} \prod_t \frac{dp\, dq}{2\pi} \tag{1.21}$$

would be achieved also if we had used another ordering procedure for the operator factors. At first sight we have succeeded in unambiguously constructing quantum mechanics entirely in terms of classical objects which are canonical invariants. Actually, this cannot be true, since the whole group of canonical transformations of classical mechanics does not act in quantum mechanics. The solution of this apparent paradox is based on the fact that, really, we have not given the definition of the path integral in internal terms without assuming the limit process.

In order to attach real meaning to the path integral, it is necessary to define the concrete way to calculate it. In doing so one must adopt a definite method for ordering non-commuting operators and also take into account the additional integration over the final momentum.

In field theory one, first of all, is interested in the limit to which the transition functions tend as $t'' \to \infty, t' \to -\infty$. In this case a rigorous definition of the path integral (and, thus, a procedure for ordering operators) is given by perturbation theory supplemented with a renormalization procedure. From this point of view, correctly defining the path integral is equivalent to constructing renormalized perturbation theory. Precisely this construction will be dealt with throughout this book.

In the meanwhile we shall deal with the path integral, treating it as if it were finite-dimentional. We hope that formal manipulation with the path integral, to be encountered below, will help the reader to develop a sufficiently clear intuitive notion of this object.

Feynman himself used a somewhat different form of the path integral, namely, the integral over the trajectories in the coordinate space. The Feynman formula is obtained if the Hamiltonian is quadratic in momenta:

$$h = \frac{p^2}{2m} + v(q). \tag{1.22}$$

Indeed, in this case the integration over the variables p can be carried out explicitly. In the integral

$$\int \exp\left\{i \int_{t'}^{t''} \left(p\dot{q} - \frac{p^2}{2m} - v(q)\right) dt\right\} \prod_t \frac{dp\, dq}{2\pi} \tag{1.23}$$

a shift

$$p(t) \to p(t) + m\dot{q}. \tag{1.24}$$

must be performed. Then the integration over p and q becomes separated, and we get the answer

$$\langle q'', t'' | q', t' \rangle = \frac{1}{N} \int \exp\left\{i \int_{t'}^{t''} \left(\frac{\dot{q}^2}{2m} - v(q)\right) dt\right\} \prod_t dq, \tag{1.25}$$

where

$$N^{-1} = \int \exp\left\{-i \int_{t'}^{t''} \frac{p^2}{2m} dt\right\} \prod_t \frac{dp}{2\pi}. \tag{1.26}$$

The normalizing factor N is obviously independent of q' and q'' and is a function only of the time $t'' - t'$. Usually this factor is included in the definition of measure. From the derivation presented above it is clear that the second form of the path integral is less general. It is correct only for Hamiltonians that are quadratic in momenta. However, for most problems that are interesting from a physical point of view the Hamiltonian possesses this property, and therefore for these problems the two forms are equivalent.

The case of a system with many degrees of freedom can be treated analogously. Using the vector notation

$$p = (p_1, \ldots, p_n), \qquad q = (q^1, \ldots, q^n),$$
$$p\dot{q} = \sum_i p_i \dot{q}^i; \qquad \frac{dp\,dq}{2\pi} = \prod_i \frac{dp_i\,dq^i}{2\pi}, \tag{1.27}$$

we can retain formula (1.21) in this case also, but formula (1.25) is modified, in the general case.

The quadratic form for the momenta of a system involving many degrees of freedom has the form

$$g^{ik}(q)p^i p^k, \tag{1.28}$$

where $g^{ik}(q)$ is the Riemann metric on the coordinate manifold.

As before, separation of integrations over p and q is performed by a shift, but the integral over p in this case has the form

$$\int \exp\left\{-i\int_{t'}^{t''} g^{ik}(q)p^i p^k\,dt\right\} \prod_i \frac{dp_i}{2\pi} = N^{-1}\prod_t (\det g)^{-1/2}, \tag{1.29}$$

where N is a constant independent of q.

The resulting analogue of the Feynman integral (1.25) assumes the following form:

$$\langle q'', t''|q', t'\rangle = \frac{1}{N}\int \exp\left\{i\int_{t'}^{t''}\left[\frac{\dot{q}_i(g^{-1})^{ik}\dot{q}_k}{2m} - v(q)\right]dt\right\} \times \prod_i (\det g)^{-1/2}\frac{dg_i}{2\pi}. \tag{1.30}$$

Unlike the case of a single variable, here there arises an additional factor $(\det g)^{-1/2}$ transforming the measure of integration over the coordinates into the Riemann measure.

From the viewpoint of the Hamiltonian dynamics the quantum field theory is a system with an infinite number of degrees of freedom. For example, in the case of a scalar neutral field described by the Lagrangian

$$\mathcal{L} = \frac{1}{2}\partial_\mu\varphi\partial_\mu\varphi - \frac{m^2}{2}\varphi^2 - V(\varphi), \tag{1.31}$$

the points in the phase space are pairs of functions $\varphi(x)$, $\pi(x)$ which form an infinite set of canonical variables. The argument x plays the role of the index number of these variables. The Poisson brackets are given by the relations:

$$\{\varphi(x), \varphi(y)\} = 0; \qquad \{\pi(x), \pi(y)\} = 0;$$
$$\{\varphi(x), \pi(y)\} = -\delta^{(3)}(x - y). \tag{1.32}$$

There exist many representations for the operators $\varphi(x)$ and $\pi(x)$ corresponding to $\varphi(x)$ and $\pi(x)$ after quantization. One representation, the coordinate one, is diagonal for $\varphi(x)$; the state vectors are functionals $\Phi(\varphi(x))$ of $\varphi(x)$, and

$$\varphi(x)\Phi(\varphi) = \varphi(x)\Phi(\varphi); \qquad \pi(x)\Phi(\varphi) = \frac{1}{i}\frac{\delta}{\delta\varphi(x)}\Phi(\varphi). \tag{1.33}$$

More often representations in the Fock space, to be introduced in Section 3, are used. The Hamiltonian corresponding to the Lagrangian (1.31) has the form

$$H(\pi,\varphi) = \int \left[\frac{\pi^2(x)}{2} + \frac{1}{2}\partial_k\varphi(x)\partial_k\varphi(x) + \frac{m^2}{2}\varphi^2(x) + V(\varphi)\right] d^3x. \tag{1.34}$$

It is easy to verify that the Hamiltonian equations of motion

$$\begin{aligned} \frac{d}{dt}\varphi(x) &= \frac{\delta H}{\delta\pi(x)} = \pi(x); \\ \frac{d}{dt}\pi(x) &= -\frac{\delta H}{\delta\varphi(x)} = \Delta\varphi - V'(\varphi) - m^2\varphi \end{aligned} \tag{1.35}$$

coincide with the usual equation for the scalar field

$$\Box\varphi + m^2\varphi = -V'(\varphi). \tag{1.36}$$

The formulas obtained, which express the evolution operator in terms of the path integral, can be directly applied to this case also. In the coordinate representation we have

$$\begin{aligned} \langle\varphi''(x),t''|\varphi'(x),t'\rangle &= \langle\varphi''(x)|\exp\{-iH(t''-t')\}|\varphi'(x)\rangle \\ &= \int \exp\left\{i\int_{t'}^{t''}\left[\pi(x,t)\partial_t\varphi(x,t) - \frac{\pi^2(x,t)}{2} - \frac{1}{2}(\partial_k\varphi(x,t))^2\right.\right. \\ &\quad \left.\left. - \frac{m^2\varphi^2(x,t)}{2} - V(\varphi(x,t))\right] d^3x\, dt\right\} \prod_{x,t}\frac{d\pi(x,t)d\varphi(x,t)}{2\pi} \\ &= \frac{1}{N}\int \exp\left\{i\int \mathcal{L}(\varphi)d^4x\right\}\prod_x d\varphi(x) \end{aligned} \tag{1.37}$$

$$t' < x_0 < t''; \qquad \varphi(x,t'') = \varphi''(x); \qquad \varphi(x,t') = \varphi'(x).$$

In the last formula we have used relativistic notation $x = (\vec{x}, t)$. The only thing in this formula that is not Lorentz invariant is the domain of integration over $t' \leq x_0 \leq t''$. Eventually we shall be interested in the evolution operator for an infinite time interval, since it is precisely this operator which is needed for construction of the scattering matrix, defined by the formula

$$S = \lim_{\substack{t''\to\infty \\ t'\to-\infty}} e^{iH_0t''}e^{-iH(t''-t')}e^{-iH_0t'}, \tag{1.38}$$

Here $H_0 - 0$ is the energy operator for free motion; it is obtained from H by omitting the interaction term $V(\varphi)$.

The representation we used before is inconvenient for calculating this limit, because the expression for the operator $\exp\{-iH_0t\}$ in this representation is rather cumbersome. A more convenient one is the so-called holomorphic representation, in which the creation operators are diagonal. The next section is devoted to the discussion of this representation.

2.2 The Path Integral in the Holomorphic Representation

We shall again begin with the case of one degree of freedom. Let us consider, as an example, a harmonic oscillator with the Hamiltonian

$$h(p,q) = \frac{p^2}{2} + \frac{\omega^2 q^2}{2}. \tag{2.1}$$

We introduce complex coordinates

$$a^* = \frac{1}{\sqrt{2\omega}}(\omega q - ip), \qquad a = \frac{1}{\sqrt{2\omega}}(\omega q + ip). \tag{2.2}$$

In terms of these coordinates the Hamiltonian has the form $h = \omega a^* a$.

In quantum mechanics, operators correspond to these coordinates. These operators are conjugates of each other and obey the commutation rules

$$[a, a^*] = 1. \tag{2.3}$$

These commutation relations have a representation in the space of analytic functions $f(a^*)$ with the scalar product

$$(f_1, f_2) = \int (f_1(a^*))^* f_2(a^*) e^{-a^* a} \frac{da^* da}{2\pi i}. \tag{2.4}$$

The operators a^* and a act in the following way:

$$a^* f(a^*) = a^* f(a^*), \qquad a f(a^*) = \frac{d}{da^*} f(a^*). \tag{2.5}$$

The introduced scalar product is positive definite. Indeed, any arbitrary analytic function $f(a^*)$ is a linear combination of the monomials

$$\Psi_n(a^*) = \frac{(a^*)^n}{\sqrt{n!}}. \tag{2.6}$$

A simple calculation shows that these monomials are orthonormalized:

$$\begin{aligned} \langle \Psi_n | \Psi_m \rangle &= \frac{1}{\sqrt{n!m!}} \int a^n (a^*)^m e^{-a^* a} \frac{da^* da}{2\pi i} \\ &= \frac{1}{\sqrt{m!n!}} \frac{1}{\pi} \int_0^\infty \rho d\rho \int_0^{2\pi} d\theta \rho^{n+m} e^{i\theta(n-m)} e^{-\rho^2} = \begin{cases} 0, & n \neq m, \\ 1, & n = m, \end{cases} \end{aligned} \tag{2.7}$$

whence there follows the positive definiteness of the scalar product. In this calculation the polar coordinates

$$a = \rho e^{i\theta}, \qquad a^* = \rho e^{-i\theta}. \tag{2.8}$$

are utilized. In this case

$$\frac{da^* da}{2\pi i} = \frac{1}{\pi}\rho d\rho d\theta. \tag{2.9}$$

It is also clear that the operators a^* and a are conjugates of each other. Indeed, taking into account that

$$a^* e^{-a^* a} = -\frac{d}{da} e^{-a^* a}; \qquad \frac{d}{da} f(a^*) = 0, \tag{2.10}$$

and integrating by parts, we have

$$\begin{aligned}(f_1, a^* f_2) &= \int (f_1(a^*))^* a^* f_2(a^*) e^{-a^* a} \frac{da^* da}{2\pi i} \\ &= \int \left[\frac{d}{da}((f_1(a^*))^* f_2(a^*))\right] e^{-a^* a} \frac{da^* da}{2\pi i} \\ &= \int \left(\frac{d}{da^*} f_1(a^*)\right)^* f_2(a^*) e^{-a^* a} \frac{da^* da}{2\pi i} = (a f_i, f_2). \end{aligned} \tag{2.11}$$

There are two ways to describe arbitrary operators in this representation. First, an arbitrary operator A can be represented by an integral operator with a kernal $A(a^*, a)$

$$(Af)(a^*) = \int A(a^*, \alpha) f(\alpha^*) e^{-\alpha^* \alpha} \frac{d\alpha^* d\alpha}{2\pi i}. \tag{2.12}$$

The kernel $A(a^*, a)$ is expressed in terms of the matrix elements of the operator A in the basis ψ_n : if

$$A_{mn} = \langle \psi_n | A | \psi_m \rangle, \tag{2.13}$$

then

$$A(a^*, a) = \sum_{n,m} A_{nm} \frac{(a^*)^n}{\sqrt{n!}} \frac{a^m}{\sqrt{m!}}. \tag{2.14}$$

This formula defines $A(a^*, a)$ as an analytic function of two complex variables, a^*, a. The functions $A(a^*, a)$ is determined uniquely by its values in the real plane, i.e., when a^* and a are complex conjugates, as in formulas (2.2) and (2.11). This condition, however, does not hold in formula (2.12) and in the subsequent formulas, and the variables a and a^* remain independent of each other.

The convolution of the kernels corresponds to the product of the operators A_1 and A_2:

$$(A_1 A_2)(a^*, a) = \int A_1(a^*, \alpha) A_2(\alpha^*, a) e^{-\alpha^* a} \frac{d\alpha^* d\alpha}{2\pi i}. \tag{2.15}$$

The second representation for operators is simply the definition of an operator in the form of a normally ordered polynomial in operators a^* and a. A product in which all operators a^* are placed to the left of all operators a is called a normal product. Let us examine the kernel of an operator A, given in terms of a sum of normal products:

$$A = \sum_{n,m} K_{nm} (a^*)^n a^m. \tag{2.16}$$

This operator can be associated with a function

$$K(a^*, a) = \sum_{n,m} K_{nm} (a^*)^n a^m, \tag{2.17}$$

We shall call this function the normal symbol of the operator A. Then the kernel $A(a^*, a)$ of the operator A is related to $K(a^*, a)$ by the formula

$$A(a^*, a) = e^{a^* a} K(a^*, a). \tag{2.18}$$

For a check we shall consider, as an operator A, the monomial

$$A = (a^*)^k a^l, \tag{2.19}$$

so that

$$K(a^*, a) = (a^*)^k a^l \tag{2.20}$$

and

$$\begin{aligned} A_{nm} &= \langle \psi_n | A | \psi_m \rangle \\ &= \frac{1}{\sqrt{n!m!}} \int \left(\left(\frac{d}{da^*} \right)^k (a^*)^n \right)^* \left(\left(\frac{d}{da^*} \right)^l (a^*)^m \right) e^{-a^* a} \frac{da^* da}{2\pi i} \\ &= \sqrt{n(n-1)\ldots(n-k+1)} \sqrt{m(m-1)\ldots(m-l+1)} \\ &\quad \times \theta(n \geq k) \theta(m \geq l) \delta_{n-k, m-l}, \end{aligned} \tag{2.21}$$

where

$$\theta(n \geq k) = \begin{cases} 0, & \text{if } n < k, \\ 1, & \text{if } n \geq k. \end{cases}$$

We shall now construct $A(a^*, a)$, using the formula (2.14). We have

$$A(a^*, a) = \sum_{n,m} A_{nm} \frac{(a^*)^n}{\sqrt{n!}} \frac{a^m}{\sqrt{m!}} = (a^*)^k a^l \sum \frac{(a^*)^n a^n}{n!} = (a^*)^k a^l \exp\{a^* a\}. \tag{2.22}$$

The formula (2.18) is thus verified. The formulas (2.18) and (2.15) allow us in a simple way to construct the evolution operator in the form of a path integral over the functions $a^*(t)$ and $a(t)$. The corresponding derivation is actually a repetition of the reasoning of Section 2.1.

Let the Hamiltonian be given in the form

$$H = h(a^*, a), \tag{2.23}$$

where a normal ordering is assumed. Then the kernal $U(a^*, a, \Delta t)$ of the evolution operator

$$U(\Delta t) = \exp\{-iH\Delta t\} \tag{2.24}$$

for small Δt takes the following form:

$$U(a^*, a, \Delta t) = \exp\{a^* a - ih(a^*, a)\Delta t\}. \tag{2.25}$$

For an arbitrary interval $t'' - t' = N\Delta t$ we must calculate the convolution of N such kernels:

$$U(a^*, a; t'' - t') = \int \exp\{[a_N^* a_{N-1} - a_N^* a_{N-1} + \ldots - a_1^* a_1 + a_1^* a_0] \\ -i[h(a_N^*, a_{N-1}) + \ldots + h(a_1^*, a_0)]\Delta t\} \prod_{k=1}^{N-1} \frac{da_k^* da_k}{2\pi i}, \tag{2.26}$$

where we have denoted $a_0 = a, a_N = a^*$. The formal list as $\Delta t \to 0$ and $N \to \infty$ is expressed by

$$U(a^*, a; t'' - t') = \int \exp\{a^*(t'')a(t'')\} \\ \times \exp\left\{\int_{t'}^{t''} (-a^*\dot{a} - ih(a^*, a))dt\right\} \prod_t \frac{da^* da}{2\pi i}, \tag{2.27}$$

or, after symmetrizing in a^* and a,

$$U(a^*, a; t'' - t') = \int \exp\left\{\frac{1}{2}(a^*(t'')a(t'') + a^*(t')a(t'))\right\} \\ \times \exp\left\{i\int_{t'}^{t''} \left[\frac{1}{2i}(\dot{a}^* a - a^*\dot{a}) - h(a^*, a)\right] dt\right\} \prod_t \frac{da^* da}{2\pi i}. \tag{2.28}$$

Here it is assumed that $a^*(t'') = a^*$ and $a(t') = a$. We may point out that the latter formula differs insignificantly from the corresponding formula (1.21) of the previous section. In both these formulas the integrand is the functional $\exp\{i\times\text{action}\}$, and the integration is carried over the product of Liouville measures over the phase space. The additional functional

$$\exp\left\{\frac{1}{2}(a^*(t'')a(t'') + a^*(t')a(t'))\right\} \tag{2.29}$$

in the formula (2.28) reflects the differences in boundary conditons on the trajectories over which we integrate: in the case of (1.21) we fix the value of the same function $q(t)$ at $t = t'$, whereas in the case of (2.27) at $t = t'$ the value of the function $a(t)$ is fixed, and $t = t''$ it is the value of the function $a^*(t)$. We emphasize that the variables $a^*(t'')$ and $a(t'')$ are independent; we integrate over $a(t'')$, but $a^*(t'')$ remains fixed. Analogously, we integrate over the variable $a^*(t')$, leaving $a(t')$ fixed.

In the case of a harmonic oscillator the integral (2.28) is easily computed, since the integrand is an exponential function of a nonuniform quadratic form.

We shall call such integrals Gaussian integrals. We shall take advantage of the well-known property of a Gaussian integral, according to which it is equal (up to a constant that will be systematically included in the definition of the integration measure) to the value of the integrand, calculated at the extremum point of the exponent of the exponential function. The extremum, in this case, does not necessarily lie within the region of integration. The condition for the extremum in our case coincides with the classical equation of motion

$$\dot{a}^* - i\omega a^* = 0; \dot{a} + i\omega a = 0; a^*(t'') = a^*; a(t') = a, \tag{2.30}$$

since

$$\delta(a^*(t'')a(t'') + \int_{t'}^{t''} (-a^*\dot{a} - i\omega a^* a)dt) = \int_{t'}^{t''} (\delta a(\dot{a}^* - i\omega a^*) - \delta a^*(\dot{a} + i\omega a))dt \tag{2.31}$$

at $\delta a^*|_{t''} = 0; \delta a|_{t'} = 0$.

Equations (2.30) can be solved in a trivial manner:

$$a(t) = e^{i\omega(t-t'')}a; \qquad a^*(t) = e^{i\omega(t-t'')}a^*. \tag{2.32}$$

When the solutions (2.32) are substituted into the action taking part in formula (2.27), a non-trivial contribution is given only by the term outside of the integral. As a result, for the kernel of the evolution operator, which will be denoted by U_0, we obtain

$$U_0(a^*, a; t'' - t') = \exp\{a^* a e^{i\omega(t'-t'')}\}. \tag{2.33}$$

If $f(a^*)$ is an arbitrary function, then

$$U_0(t)f(a^*) = \int \exp\{a^* \alpha e^{-i\omega t}\} f(\alpha^*) e^{-\alpha^* \alpha} \frac{d\alpha^* d\alpha}{2\pi i} = f(a^* e^{-i\omega t}). \tag{2.34}$$

This formula clearly demonstrates the convenience of using the holomorphic representation for a harmonic oscillator. In this representation the evolution of an arbitrary state $\Phi(a^*)$ is reduced to substitution of the argument,

$$a^* \to a^* e^{-i\omega t}. \tag{2.35}$$

This property is very useful for the field theory, because in this case the free Hamiltonian is represented by the sum of the Hamiltonians of an infinite set of oscillators.

2.3 The Generating Functional for the S-Matrix in Field Theory

The holomorphic representation is introduced in the field theory in terms of complex amplitudes $a^*(k)$ and $a(k)$. The canonical variables $\varphi(x)$ and $\pi(x)$ for a scalar field are expressed in terms of these amplitudes in the following way:

$$\begin{aligned}\varphi(x) &= \left(\frac{1}{2\pi}\right)^{3/2} \int (a^*(k)e^{-ikx} + a(k)e^{ikx})\frac{d^3k}{\sqrt{2\omega}}, \\ \pi(x) &= \left(\frac{1}{2\pi}\right)^{3/2} \int (a^*(k)e^{-ikx} - a(k)e^{ikx})i\sqrt{\frac{\omega}{2}}d^3k,\end{aligned} \tag{3.1}$$

$$k_0 = \omega = (k + m^2)^{1/2}.$$

The free Hamiltonian H_0 is expressed in terms of a^* and a as follows:

$$H_0 = \int \omega(k)a^*(k)a(k)d^3k, \tag{3.2}$$

and is the sum of the energies of an infinite set of oscillators. The argument k plays the role of the oscillator number, and $\omega(k)$ plays the role of its frequency.

Under quantization the amplitudes $a^*(k)$ and $a(k)$ acquire the meaning of creation and annihilation operators, respectively. Such terminology is due to the fact that the state Φ, given in the holomorphic representation by the functional

$$\Phi_n(a^*) = \prod_{i=1}^{n} a^*(k_i), \tag{3.3}$$

is an eigenstate of the Hamiltonian (3.2) with the eigenvalue $\sum_{i=1}^{n} \omega(k_i)$. This state is interpreted as a state of n free particles with momenta $k_1, \ldots, k_n$ and is denoted by the symbol $|k_1, \ldots, k_n\rangle$. The state without particles, $|0\rangle$, is represented by a functional identically equal to unity. An arbitrary n-particle state is a linear combination of these states,

$$\sum_n \int c_n(k_1, \ldots, k_n)|k_1, \ldots, k_n\rangle \prod_i dk_i, \tag{3.4}$$

where the functions c_n are symmetric with respect to their arguments.

The operators a^* and a transform an n-particle state into a state with $n+1$ particles and a state with $n+1$ particles into an n-particle state, respectively. Their action on the basis functions $k_1, \ldots, k_n$ is of the following form:

$$a^*(k)|k_1,\dots,k_n\rangle = |k_1,\dots,k_n,k\rangle, \tag{3.5}$$

$$a(k)|k_1,\dots,k_n\rangle = \sum_{i=1}^{n} \delta(k-k_i)|k_1,\dots,k_{i-1},k_{i+1},\dots,k_n\rangle,$$

$$a(k)|0\rangle = 0. \tag{3.6}$$

The space constituted by linear combinations (3.4) is conventionally known as the Fock space. A state vector is defined in it by a set of amplitudes $c = \{c_n(k_1,\dots,k_n)\}$. A scalar product is given by the formula

$$(c',c'') = \sum_n n! \int c_n^{*\prime}(k_1,\dots,k_n)c_n''(k_1,\dots,k_n)\prod_i dk_i. \tag{3.7}$$

The complete Hamiltonian H, besides the term H_0, contains the interaction $V(a^*,a)$, which is obtained by substitution of the function $\varphi(x)$ in the form (3.1) into $\int V(\varphi)d^3x$. The evolution operator $U(t'',t')$ is determined by the kernel $U(a^*(k), a(k), t''-t')$, which is expressed in terms of the path integral

$$U(a^*(k),a(k),t''-t') = \int \exp\Big\{\int d^3k a^*(k,t'')a(k,t'') + \int_{t'}^{t''}\Big[-iV(a^*,a) + \int d^3k(-a^*(k,t)\dot{a}(k,t) - i\omega a^*(k,t)a(k,t))\Big]dt\Big\} \times \prod_{t,k}\frac{da^*(k,t)da(k,t)}{2\pi i};$$

$$a^*(k,t'') = a^*(k); \qquad a(k,t') = a(k). \tag{3.8}$$

From this formula it is easy to pass over to the S-matrix. For this we point out that for an arbitrary operator A with a kernel $A(a^*(k),a(k))$ the operator

$$e^{iH_0t''}Ae^{-iH_0t'} \tag{3.9}$$

has the kernel

$$A(a^*(k)e^{i\omega t''}, a(k)e^{-i\omega t'}). \tag{3.10}$$

This is a direct generalization of the formula (2.33) for a harmonic oscillator, obtained in the previous section. Thus, the kernel of the S-matrix is obtained as the limit for $t'' \to \infty$ and $t' \to \infty$ of the integral (3.3), which we shall for convenience rewrite in a symmetrized form as

$$S(a^*(k),a(k)) = \lim_{\substack{t''\to\infty \\ t'\to-\infty}} \int \exp\Big\{\frac{1}{2}\int d^3k(a^*(k,t'')a(k,t'') + a^*(k,t')a(k,t')) + i\int_{t'}^{t''} dt\Big[\int d^3k\Big(\frac{1}{2i}(\dot{a}^*(k,t)a(k,t) - a^*(k,t)\dot{a}(k,t)) - \omega(k)a^*(k,t)a(k,t)\Big) - V(a^*,a)\Big]\Big\}\prod_{k,t}\frac{da^*(k,t)da(k,t)}{2\pi i} \tag{3.11}$$

where

$$a^*(k,t'') = a^*(k)\exp\{i\omega(k)t''\}, \tag{3.12}$$

$$a(k,t') = a(k)\exp\{-i\omega(k)t'\}. \tag{3.13}$$

We shall apply this formula to calculating the S-matrix for scattering on an external source $\eta(x)$, when

$$V(\varphi) = -\eta(x)\varphi(x) \tag{3.14}$$

and $\eta(x)$ decreases rapidly as $|t| \to \infty$.

The corresponding functional $V(a^a, a)$ has the form

$$V(a^*,a) = \int d^3k[\gamma(k,t)a^*(k) + \gamma^*(k,t)a(k)], \tag{3.15}$$

where

$$\gamma(k,t) = -\frac{1}{\sqrt{2k_0}}\left(\frac{1}{2\pi}\right)^{3/2}\int \eta(x,t)e^{-ikx}d^3x. \tag{3.16}$$

The functional $V(a^*, a)$ depends explicitly on time. Nevertheless, all the formulas for the evolution operator in this case still remain valid. The only change is that the evolution operator now depends on both variables t'' and t' and not only on the difference between them. The integrand in (3.11) in our case again takes the form of an exponential function of a nonuniform quadratic form, and the Gaussian integral is calculated in the same manner as in the previous section. The conditions for the extremum are the following:

$$\begin{gathered}\dot{a}(k,t) + i\omega(k)a(k,t) + i\gamma(k,t) = 0, \\ \dot{a}^*(k,t) - i\omega(k)a^*(k,t) - i\gamma^*(k,t) = 0, \\ a^*(k,t'') = a^*(k)e^{i\omega t''}, \quad a(k,t') = a(k)e^{-i\omega t'}.\end{gathered} \tag{3.17}$$

The solution of these equations is given by the formulas

$$a^*(k,t) = a^*(k)e^{i\omega t} - ie^{i\omega t}\int_t^{t''} e^{-i\omega s}\gamma^*(k,s)ds, \tag{3.18}$$

$$a(k,t) = a(k)e^{-i\omega t} - ie^{-i\omega t}\int_{t'}^{t} e^{i\omega s}\gamma(k,s)ds. \tag{3.19}$$

Substituting this solution into the exponential function in the formula (3.11) and passing to the limit, we obtain the following expression for the kernel of the S-matrix:

$$S_\eta(a^*,a)=\exp\Big\{\int a^3k\Big[a^*(k)a(k)$$
$$+\frac{1}{(2\pi)^{3/2}}\int_{-\infty}^{\infty}dt\int d^3x\eta(x,t)\frac{a^*(k)e^{i\omega t}e^{-i(kx)}+a(k)e^{-i\omega t}e^{i(kx)}}{\sqrt{2\omega}}$$
$$-\left(\frac{1}{2\pi}\right)^3\frac{1}{2}\int_{-\infty}^{\infty}dt\int_{-\infty}^{\infty}ds\int d^3x\int d^3y\frac{1}{2\omega}\eta(x,t)$$
$$\times\ \eta(y,s)e^{ik(x-y)}e^{-i\omega|t-s|}\Big]\Big\}. \tag{3.20}$$

The expression for the S-matrix becomes more elegant if we pass from the kernel to the normal symbol, which is equivalent to omitting the first factor $\exp\{d^3k\cdot a^*(k)a(k)\}$. The remaining factors may be rewritten in a manifastly relativistic-invariant form. For this we introduce the solution of the free Klein-Gordon equation

$$\varphi_0(x)=\left(\frac{1}{2\pi}\right)^{3/2}\int(a^*(k)e^{ikx}+a(k)e^{-ikx})\frac{d^3k}{\sqrt{2k_0}},\qquad k_0=\omega, \tag{3.21}$$

$$\Box\varphi_0+m^2\varphi_0=0 \tag{3.22}$$

and the Green function of this equation

$$\begin{aligned}D_c(x)&=-\left(\frac{1}{2\pi}\right)^3\int e^{ikx}e^{-i\omega|x_0|}\frac{d^3k}{2\omega}\\&=-\left(\frac{1}{2\pi}\right)^4\int e^{-ikx}\frac{1}{k^2-m^2+i0}d^4k,\end{aligned} \tag{3.23}$$

$$(\Box+m^2)D_c=\delta^4(x). \tag{3.24}$$

The first representation for D_c follows from the second one upon integration over k_0.

The normal symbol of the S-matrix $S(a^*,a)$ in the above notation is given by the formula

$$S_\eta(a^*,a)=\exp\left\{i\int\eta(x)\varphi_0(x)dx+\frac{i}{2}\int\eta(x)D_c(x-y)\eta(y)dx\,dy\right\}. \tag{3.25}$$

It must be stressed that taking into account the boundary conditions (3.12) and (3.13) on the trajectory of integration has led to the Green function of the Klein-Gordon equation operator $\Box+m^2$ being chosen in a unique manner in the form (3.23). This Green function is conventionally called causal.

Now let us pass to the consideration of the S-matrix in the case of the general potential $V(\varphi)$. In this case we evidently cannot calculate the corresponding path integral exactly, and we shall restrict ourselves to the construction of a perturbation theory for it. We shall show that in this case the problem is reduced to the already solved problem concerning the scattering on an external field. For this purpose we shall make use of the obvious formula

$$\varphi(x_1)\dots\varphi(x_n) = \frac{1}{i}\frac{\delta}{\delta\eta(x_1)}\dots\frac{1}{i}\frac{\delta}{\delta\eta(x_n)}\exp\left\{i\int\varphi(x)\eta(x)dx\right\}\Big|_{\eta=0}, \quad (3.26)$$

From this formula it follows that an arbitrary functional $\Phi(\varphi)$ of $\varphi(x)$ can be written in the form

$$\Phi(\varphi) = \Phi\left(\frac{1}{i}\frac{\delta}{\delta\eta(x)}\right)\exp\left\{i\int\varphi(x)\eta(x)dx\right\}\Big|_{\eta=0} \quad (3.27)$$

In particular,

$$\exp\left\{-i\int V(\varphi)dx\right\} = \exp\left\{-i\int V\left(\frac{1}{i}\frac{\delta}{\delta\eta}\right)dx\right\}\exp\left\{i\int\varphi\eta\,dx\right\}\Big|_{\eta=0}. \quad (3.28)$$

This formula, of course, is understood in the sense of perturbation theory.

Thus, in the path integral (3.11), which determines the S-matrix for a potential of the general form, we can substitute the $\exp\{-i\int V(\varphi)dx\}$ in the integrand by the right-hand side of (3.28) and put outside of the path integral the formal differential operator

$$\exp\left\{-i\int V\left(\frac{1}{i}\frac{\delta}{\delta\eta}\right)dx\right\}$$

The remaining path integral coincides precisely with the already calculated integral for the S-matrix for scattering on an external source. As a result, we obtain the following final expression for the normal symbol of the S-matrix:

$$S(a^*,a) = S(\varphi_0) = \exp\left\{-i\int V\left(\frac{1}{i}\frac{\delta}{\delta\eta(x)}\right)dx\right\}$$
$$\exp\left\{i\int\eta(x)\varphi_0(x)dx + \frac{i}{2}\int\eta(x)D_c(x-y)\eta(y)dx\,dy\right\}\Big|_{\eta=0}. \quad (3.29)$$

Here we have replaced the pair of arguments a^*, a by a single function φ_0, since they define each other uniquely. Expanding this functional in a series in φ_0,

$$S(\varphi_0) = \sum_n \frac{1}{n!}\int S_n(x_1,\dots,x_n)\varphi_0(x_1)\dots\varphi_0(x_n)dx_1\dots dx_n, \quad (3.30)$$

we obtain the coefficient functions $S_n(x_1,\dots,x_n)$. In the operator formalism these functions appear when the S-matrix operator is expanded in a series over the normal products of free fields. The functional $S(\varphi_0)$ is sometimes called the generating functional for the coefficient functions of the S-matrix.

Formula (3.29) serves as a convenient starting point for the construction of a perturbation theory for calculating the coefficient functions. We shall consider the interaction $V(\varphi)$ to be a polynomial in the fields φ, i.e., to be a linear combination of the monomials $V_n(\varphi) = \varphi^n(x)$ with small coefficients. We shall make correspond to the operator $V_n(\frac{1}{i} \cdot \frac{\delta}{\delta\eta(x)})$ a diagram (Fig. 1) in which n lines go out of the vertex indicated by the dot, x. In formula (3.29) the differential operator $\exp\left\{-i\int V\left(\frac{1}{i}\cdot\frac{\delta}{\delta\eta(x)}\right)dx\right\}$ is applied to the functional $S_\eta(\varphi_0)$ defined by the formula (3.25). Differentiating $S_\eta(\varphi_0)$ with respect to $\eta(x)$ we obtain

$$\frac{\delta}{\delta\eta(x)}S_\eta(\varphi_0) = \left(i\varphi_0(x) + i\int D_c(c-y)\eta(y)dy\right)S_\eta(\varphi_0). \tag{3.31}$$

Calculation of $S(\varphi_0)$ by formula (3.29) reduces to multiple differentiation of $S_\eta(\varphi_0)$ with respect to $\eta(x_i)$ with subsequent integration over x_i. Here, the second summand in (3.31) yields a non-trivial contribution for $\eta = 0$ only if it is idfferentiated once more. Clearly, multiple application of the operator $\frac{\delta}{\delta\eta(x_i)}$ to $S_\eta(\varphi_0)$ will lead, for $\eta = 0$, to the appearance of the factors $\varphi_0(x_i)$ or $D_c(x_i - x_j)$.

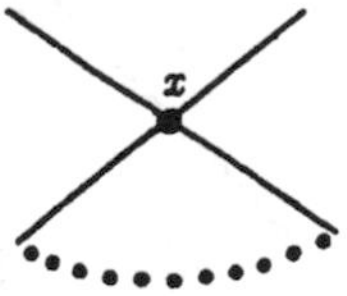

Figure 1.

In accordance with the above the function $D_c(x_i - x_j)$ corresponds to the second-order differential operator $\frac{\delta^2}{\delta\eta(x_i)\delta\eta(x_j)}$. We can now readily depict graphically an arbitrary term of the expansion $S(\varphi_0)$ in V_n. We shall make a diagram consisting of k vertices, indicated by the points $x_i (i = 1, \dots, k)$ from each one of which n_1 lines depart, corresponding to the product of k differential operators $V_{n_1}(\frac{1}{i}\cdot\frac{\delta}{\delta\eta(x_i)}), i = 1, \dots, k$. These lines may either be closed, connect a pair of vertices, (we shall call them internal), or go out of the diagram (we shall call them external). A function $D_c(x_i - x_j)$ is made to correspond to each internal line connecting the points x_i and x_j; a field $\varphi_0(x_i)$ is made to correspond to each external line leaving the point x_i. The total expression for $S(\varphi_0)$ is obtained by summation over all possible diagrams with subsequent integration over x_i.

As an example, we shall consider the most simple interaction

$$V(\varphi) = \frac{g}{3!}\varphi^3(x), \tag{3.32}$$

where g is a small parameter called the coupling constant. In this case there exists only one sort of vertices with three out-going lines (Fig. 2).

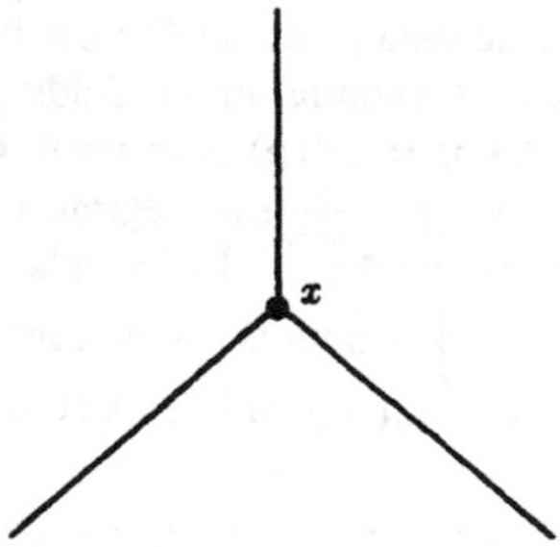

Figure 2.

In the lowest order in the coupling constant $S(\varphi_0)$ is described by two diagrams (Fig. 3) giving contributions proportional to $\int \varphi_0^3(x)dx$ and to $D_c(0) \int \varphi_0(x)dx$, respectively.

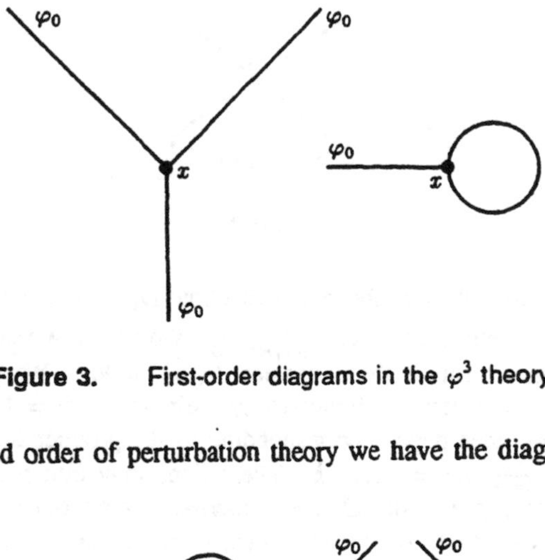

Figure 3. First-order diagrams in the φ^3 theory

In the second order of perturbation theory we have the diagrams depicted in Fig. 4.

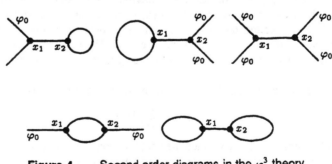

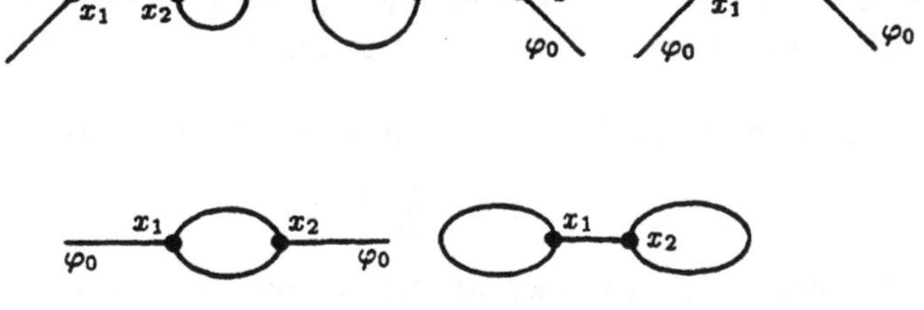

Figure 4. Second-order diagrams in the φ^3 theory

The same rules apply for calculation of these diagrams: a function $D_c(x-y)$ corresponds to each internal line, and a field φ_0 corresponds to each external line. There also arises a numerical factor, which will be discussed below.

The further construction is obvious. The described diagram technique was first developed by R. Feynman, therefore such diagrams are called Feynman diagrams (or graphs). Let us formulate the ultimate recipe for establishing the correspondence between $S-n(\varphi_0)$ and Feynman diagrams. One must draw n vertices indicated by dots, $(x_1,\ldots,x_n)$, like the one in Fig. 2, and then, closing all the possible solid lines, we obtain a sum of the diagrams depicted in Fig. 3, Fig. 4, and so on.

A field $\varphi_0(x_i)$ must be made to correspond to each external line leaving the point x_i; a Green function (or propagator), $iD_c(x_i-x_j)$, must be made to correspond to each internal line connecting the points x_i and x_j. The expression thus obtained must be multiplied by $(-ig)^n \cdot \frac{1}{(3!)^n n!}$ and integrated over all x_i.

We note, however, that when proceeding in this manner we in no way took into account the symmetry of the diagrams, so many of the diagrams thus obtained will differ only by trivial redefinition of integration variables. To take this fact into account, we note that n independent vertices of the type depicted in Fig. 2 exhibit group symmetry of the order $n!(3!)^n$. The factor $(n!)$ reflects symmetry with respect to the permutation of n equivalent vertices, while the factor $(3!)^n$ reflects symmetry relative to rearrangement of the lines in the vertices. Therefore for obtaining an analytical expression for $(S(\varphi_0))_n$ it suffices to depict only essentially different diagrams, having arranged the arguments $x_1,\ldots,x_n$ arbitrarily, and to multiply the result by the number of ways the given diagram can be obtained from the given vertices. This number equals $n!(3!)^n/D_n$, where D_n is the order of the symmetry group of the diagram. As a result of such multiplication, the factor $(n!)(3!)^n$ cancels out with the similar factor before the integral and the entire combinatorial factor reduces to $(D_n)^{-1}$.

The diagrams obtained in the above way will, in particular, contain disconnected diagrams such as, for instance, the ones depicted in Fig. 5.

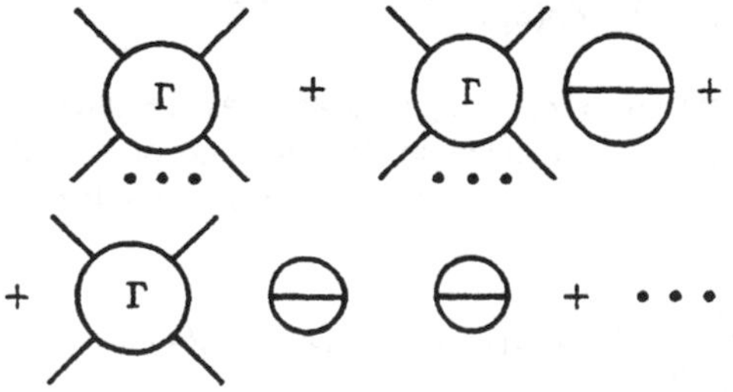

Figure 5. Factorization of vacuum diagrams in the φ^3 theory

Individual terms in this sum differ from each other by factors representing sums of all possile diagrams without external lines (vacuum diagrams). Each such term, obviously, exhibits a symmetry of order $m!$, where m is the number of identical connected components of vacuum diagrams. Therefore an arbitrary diagram Γ containing no disconnected vacuum components is multiplied by the series

$$\left(\sum_{m=0}^{\infty} \frac{(\Gamma_0)^m}{m!}\right) = e^{\Gamma_0} \tag{3.33}$$

Here Γ_0 is the sum of all the connected vacuum diagrams. Since a similar constant factor appears before all diagrams, we can include it in the definition of the integral measure and redefine the normal symbol of the S-matrix by dividing it by the sum of vacuum diagrams

$$\tilde{S}(\varphi_0) = \frac{S(\varphi_0)}{e^{\Gamma_0}} = \frac{S(\varphi_0)}{S(0)}. \tag{3.34}$$

The expansion of $\tilde{S}(\varphi_0)$ in a perturbation series no longer contains vacuum diagrams.

The described diagram technique is readily generalized to other models of field theory. Then, in practical computations the diagram technique in the momentum representation (when φ_0 and D are replaced by their Fourier transforms, $\tilde{\varphi}_0(p)$ amd $\tilde{D}(p)$) is usually applied. In the next chapter we shall formulate the Feynman rules for the Yang-Mills field. If in the formula (3.29) we do not assume $\eta = 0$, then the resulting functional

$$S(\varphi_0, \eta) = \exp\left\{-i \int V\left(\frac{1}{i}\frac{\delta}{\delta\eta(x)}\right) dx \exp\left\{i \int \eta(x)\varphi_0(x)dx\right\}\right.$$
$$\left. \times \exp\left\{\frac{i}{2}\int \eta(x)D_c(x-y)\eta(y)dx\, dy\right\}\right\} \tag{3.35}$$

is the normal symbol of the S-matrix for the scattering of interacting particles in the presence of an external source $\eta(x)$. In practice, it is often more convenient to deal not with the S-matrix (3.29), but with the functional

$$Z(\eta) = \exp\left\{-i \int V\left(\frac{1}{i}\frac{\delta}{\delta\eta(x)}\right) dx\right\} \times \exp\left\{\frac{i}{2}\int \eta(x)D_c(x-y)\eta(y)dx\, dy\right\}, \tag{3.36}$$

which coincides with $S(\varphi_0, \eta)$ at $\varphi_0 = 0$ and has the meaning of the transition amplitude from vacuum to vacuum in the presence of an external source. The coefficient functions $G_n(x_1, \ldots, x_n)$ in the expansion of this functional in the series $\eta(x)$,

$$Z(\eta) = \sum_n \frac{1}{n!}\int G_n(x_1, \ldots, x_n)\eta(x_1)\ldots\eta(x_n)dx_1\ldots dx_n \tag{3.37}$$

define the so-called Green functions, to which in the operator formalism the average values of the chronological products of the Heisenberg field operators correspond. The Green functions are necessary, in particular, for realization of the renormalization program, which will be discussed in the next chapters and which until now has not yet been formulated for the S-matrix directly.

The functional $Z(\eta)$ itself contains more information than $S(\varphi_0)$, since it is defined for arbitrary functions η, whereas $S(\varphi_0)$ is defined only on the mass shell, that is, its argument φ_0 is the solution of the free-field equation of motion. Knowledge of the functional $Z(\eta)$ allows the reconstruction of $S(\varphi_0)$. The corresponding procedure is determined by the so-called reduction formulas, which are readily deduced from comparison of the formulas (3.29) and (3.36).

In order to obtain explicit formulas, we introduce the extended functional $\tilde{S}(\varphi)$ by replacing φ_0 in (3.29) with an arbitrary function of four variables. Then the coefficient functions extended off the mass shell are variational derivatives.

$$\tilde{S}_n(x_1, \ldots, x_n) = \frac{1}{i}\frac{\delta}{\delta\varphi(x_1)} \cdots \frac{1}{i}\frac{\delta}{\delta\varphi(x_n)}\tilde{S}(\varphi)|_{\varphi=0}.$$

On the other hand, we can replace the argument $\eta(x)$ in the functional $Z(\eta)$ by $\bar{\eta}(x)$, where

$$\bar{\eta}(x) = \int D_c(x-y)\eta(y)\,dy. \tag{3.38}$$

Then by direct comparison we verify that

$$\int \prod_i dx_i \varphi_0(x_i) \left\{ \frac{1}{i}\frac{\delta}{\delta\varphi(x_1)} \cdots \frac{1}{i}\frac{\delta}{\delta\varphi(x_n)}\tilde{S}(\varphi)|_{\varphi=0} - \frac{1}{i}\frac{\delta}{\delta\bar{\eta}(x_1)} \cdots \frac{1}{i}\frac{\delta}{\delta\bar{\eta}(x_n)} Z(\bar{\eta})|_{\eta=0} \right\} = 0. \tag{3.39}$$

Thus, we have a simple procedure for calculating the normal symbol of the S-matrix. One must calculate the variational derivatives

$$\frac{1}{i}\frac{\delta}{\delta\eta(x_1)} \cdots \frac{1}{i}\frac{\delta}{\delta\eta(x_n)} Z(\eta)|_{\eta=0}, \tag{3.40}$$

(that is, the Green functions $G_n(x_1, \ldots, x_n)$), apply to these functions the differential operator

$$\prod_{i=1}^{n} (\Box_{x_i} + m^2), \tag{3.41}$$

then multiply the result by the product

$$\frac{1}{n!}\prod_i \varphi_0(x_i), \tag{3.42}$$

integrate over all x, and sum over n.

2.4 The S-matrix as a Functional on Classical Solutions

In this section we shall demonstrate that expression (3.29) for the S-matrix obtained above and including the formal differential operator $\exp\left\{-i \int V\left(\frac{1}{i}\frac{\delta}{\delta\eta}\right) dx\right\}$

(Lorentz) can be written in the form of the path integral of a manifestly relativistically invariant functional. To this end we make use of the formula

$$\exp\left\{\frac{i}{2}\int \eta(x)D_c(x-y)\eta(y)dx\,dy\right\}$$
$$= \int \prod_x d\alpha(x) \exp\left\{-\int \frac{i}{2}\alpha(x)D_c^{-1}(x-y)\alpha(y)dx\,dy + i\int \alpha(x)\eta(x)dx\right\}. \quad (4.1)$$

The distribution $D_c^{-1}(x-y)$ can be written as a result of the application of the Klein-Gordon operator to the function $\delta^4(x-y)$.

Here the functions $\alpha(x)$ must belong to the class of functions on which the Klein-Gordon operator is uniquely reversible, while its inverse operator is given by the kernel $D_c(x-y)$. In other wordls, the functions $\alpha(x)$ must permit a representation of the form

$$\alpha(x) = \int D_c(x-y)\beta(y)\,dy, \quad (4.2)$$

where $\beta(x)$ falls off rapidly as $|t| \to \infty$.

Formula (4.1) actually represents the definition of a Gaussian integral, in accordance with which it is determined by the value of the integrand at the point of the extremum.

More explicitly, the boundary condition (4.2) means that the functions $\alpha(x)$ coincide asymptotically with the solutions of the free Klein-Gordon equation; as $|t| \to \infty$:

$$\alpha_-(x) = \frac{1}{(2\pi)^{3/2}}\int b_-(k)e^{ikx}|_{k_0=\omega}d^3k \quad \text{as } t \to -\infty, \quad (4.3)$$

$$\alpha_+(x) = \frac{1}{(2\pi)^{3/2}}\int b_+(k)e^{-ikx}|_{k_0=\omega}d^3k \quad \text{as } t \to +\infty, \quad (4.4)$$

i.e., they contain only the positive frequency exponential $e^{i\omega t}$ when $t \to -\infty$, and the negative frequency exponential $e^{-i\omega t}$ when $t \to +\infty$. We shall term the described asymptotic conditions the Feynman conditions, or the radiation conditions.

Substituting formula (4.1) into expression (3.29) for the S-matrix, we obtain

$$S(\varphi_0) = \exp\left\{-i\int V\left(\frac{1}{i}\frac{\delta}{\delta\eta(x)}\right)dx\right\}\int\prod_x d\alpha(x)$$
$$\times \exp\left\{i\int\left[\frac{1}{2}\partial_\mu\alpha(x)\partial_\mu\alpha(x) - \frac{m^2}{2}\alpha^2(x) + (\varphi_0(x)+\alpha(x))\eta(x)\right]dx\right\}\Bigg|_{\eta=0}. \quad (4.5)$$

Differentiation with respect to η can be performed explicitly (compare (3.28)). As a result, we obtain for $S(\varphi_0)$ the expression

$$S(\varphi_0) = \int \exp\left\{i\int\left[\frac{1}{2}\partial_\mu\alpha(x)\partial_\mu\alpha(x) - \frac{m^2}{2}\alpha^2(x) + V(\varphi_0+\alpha)\right]dx\right\}\prod_x d\alpha(x). \quad (4.6)$$

This formula may be rewritten in a more clear manner by passing to new integration variables:

$$\varphi(x) = \varphi_0(x) + \alpha(x). \tag{4.7}$$

In terms of these variables $S(\varphi_0)$ can be written in the form

$$S(\varphi_0) = \int \exp\left\{i \int \left[\frac{1}{2}\partial_\mu \varphi(x)\partial_\mu \varphi(x) - \frac{m^2}{2}\varphi^2(x) + V(\varphi)\right] dx\right\} \prod_x d\varphi(x), \tag{4.8}$$

i.e., in the form of a path integral of the function $\exp\{i \int \mathcal{L}(\varphi(x))dx\}$, if the following regularization is applied in the action quadratic form:

$$\int (\partial_\mu \varphi \partial_\mu \varphi - m^2\varphi^2)\, dx = \int [\partial_\mu(\varphi - \varphi_0)\partial_\mu(\varphi - \varphi_0) - m^2(\varphi - \varphi_0)^2]\, dx. \tag{4.9}$$

The left-hand side can formally be transformed into the right-hand side, if in the course of integration by parts the boundary terms are dropped. In this case the class of functions $\varphi(x)$ over which integration is performed is described as follows: As $t \to -\infty$ and $t \to +\infty$, $\varphi(x)$ coincides asymptotically with the solution of the free equation, $\varphi_{\text{in}}(x)$ and $\varphi_{\text{out}}(x)$, where

$$\varphi_{\text{in}}(x) = \varphi_0(x) + \alpha_-(x), \tag{4.10}$$

$$\varphi_{\text{out}}(x) = \varphi_0(x) + \alpha_+(x). \tag{4.11}$$

In other words, the positive frequency wave in φ_{in} and the negative frequency wave in φ_{out} coincide with the corresponding waves in φ_0.

As a result, we have obtained for the normal symbol ofthe S-matrix the formula

$$S(\varphi_0) = N^{-1} \int\limits_{\substack{\varphi \to \varphi_{\text{in}} \\ \varphi \to \varphi_{\text{out}}}} \exp\left\{i \int \mathcal{L}(x)\, dx\right\} \prod_x d\varphi(x). \tag{4.12}$$

We shall use this formula for deriving the alternative diagram expansion for $S(\varphi_0)$. To this end we apply the method of stationary phase directly to the integral (4.8). The stationary point φ_{cl} is determined by the solution of the classical equation of motion

$$(\Box + m^2)\varphi_{\text{cl}} = V'(\varphi_{\text{cl}}), \tag{4.13}$$

where V' stands for the derivative of the function $V(\varphi)$ with respect to the argument satisfying the asymptotic conditions

$$\varphi_{\text{cl}}(x) \to \begin{matrix} \varphi_{\text{in}}, & t \to -\infty, \\ \varphi_{\text{out}}, & t \to +\infty. \end{matrix} \tag{4.14}$$

This differential equation can be combined together with the boundary conditions in a single integral equation:

$$\varphi_{\text{cl}}(x) = \varphi_0(x) + \int D_c(x - y)V'(\varphi_{\text{cl}})(y)dy. \tag{4.15}$$

To construct the stationary phase expansion, we represent the variables of integration in the form

$$\varphi(x) = \varphi_{\rm cl}(x) + \varphi_1(x) \tag{4.16}$$

and assume the deviation of $\varphi_1(x)$ to be small.

Owing to the asymptotic conditions formulated above, $\varphi_1(x)$ satisfies the radiation conditions and, consequently, pertains to the region where the operator D_c^{-1} is defined.

The action functional is written in the form

$$\int \mathcal{L}(x)dx = \int \mathcal{L}(\varphi_{\rm cl}(x))dx + \frac{1}{2}\int [\partial_\mu\varphi_1\partial_\mu\varphi_1 - m^2\varphi_1^2 + V''(\varphi_{\rm cl})\varphi_1^2]dx$$
$$+\frac{1}{3!}\int V'''(\varphi_{\rm cl})\varphi_1^3 dx + \ldots, \tag{4.17}$$

where the dots stand for the consecutive terms of the expansion of $V(\varphi)$ in a Taylor series in the vicinity of the point $\varphi_{\rm cl}$. The quadratic form occurring in the action in the first term of the right-hand side is to be understood inthe sense of (4.9).

Substituting the expansion (4.17) into the expression for the S-matrix (4.12), we represent it in the form

$$S(\varphi_0) = S_{\rm cl}(\varphi_0) \cdot S_{\rm loop}, \tag{4.18}$$

where

$$S_{\rm cl}(\varphi_0) \exp\{i\int \mathcal{L}(\varphi_{\rm cl}(x))dx\} \tag{4.19}$$

involving the regularization of the quadratic form already mentioned, and $S_{\rm loop}$ is determined by the sum of vacuum diagrams, for the construction of which the following elements are utilized. The vertices with n outgoing lines have as coefficients the functions $1/n!V^{(n)}(\varphi_{\rm cl}(x))$, while the Green functions of the differential operator

$$\Box + m^2 - V''(\varphi_{\rm cl})_0 \equiv K(\varphi_{\rm cl}) \tag{4.20}$$

correspond to the internal lines. Besides this, a contribution to $S_{\rm loop}$ is provided by the factor $(\det K(\varphi_{\rm cl}))^{-1/2}$ which appears when the Gaussian integral over φ_1 is computed. Earlier we omitted the corresponding factor and included it in the normalizing constant. In our case, however, this cannot be done, since it depends on $\varphi_{\rm cl}$ and gives a nontrivial contribution to the S-matrix.

Ordinary Feynman diagrams arise if in the described procedure the perturbation-theory expansion of the solution of equation (4.15) is substituted for $\varphi_{\rm cl}$. In this case $S_{\rm cl}(\varphi_0)$ corresponds to the sum of tree diagrams, and $[\det K(\varphi_{\rm cl})]^{-1/2}$ gives the sum of one-loop diagrams, while the successive terms correspond to diagrams with two or more loops.

The expansion described above is known as the expansion in loops. Since the closed expression for $\varphi_{\rm cl}$ is unknown in the general case, this expansion is applied mainly in theoretical analysis, but not for concrete calculations.

Making use of (4.12) we can also immediately write out the expression for the generating functional of the Green functions in terms of the path integral. Since $Z(\eta)$ is the transition amplitude from vacuum to vacuum in the presence of the source $\eta(x)$, we have

$$Z(\eta) = N^{-1} \int \exp\left\{ i \int [\mathcal{L}(x) + \eta(x)\varphi(x)]dx \right\} \prod_x d\varphi(x), \tag{4.21}$$

where integration is performed over the fields $\varphi(x)$ satisfying the radiation condition.

The obtained formulas (4.12) and (4.21) are appealing owing to their being of a compact and clear form. Thus, representation of $Z(\eta)$ in the form of an integral permits application of simple formulas of integration, exchanging variables, computation by the method of stationary phase. Regretfully, as already mentioned above, at present there exists no definition of this integral in consistent internal terms, which would render these formal transformations strict. Nevertheless, in the framework of perturbation theory a rigorous meaning can be assigned to formula (4.21) using formula (3.36), which expresses $Z(\eta)$ in terms of variational derivatives.

On the basis of this formula one can rigorously lay the foundation of all the operations listed above for the integral (4.21) within the framework of perturbation theory. This will be done in Section 2.6.

2.5 The Path Integral over Fermi Fields

The technique described in the previous sections can be applied practically without any changes to the case of several interacting scalar fields, and also to other Bose fields, including vector fields, which will be discussed in detail in the next chapter. In the present section we shall show that for Fermi fields it is possible to construct such an integration procedure, and that the corresponding dynamic formulas (the evolution operator, the S-matrix) will look practically the same as in the case of Bose fields.

Let us start with a Fermi system having one degree of freedom. The space of states of such a system is two-dimensional. In this space two operators a^* and a act. These two operators are complex conjugates of each other and satisfy commutation rules

$$a^*a + aa^* = 1, \quad (a^*)^2 = 0; \quad a^2 = 0. \tag{5.1}$$

These operators may be represented by 2×2 matrices

$$a^* = \begin{pmatrix} 0 & 1 \\ 0 & 0 \end{pmatrix}, \quad a = \begin{pmatrix} 0 & 0 \\ 1 & 0 \end{pmatrix}. \tag{5.2}$$

The formalism of path integration is based on another representation of the operators a^*, a, which is a particular analog of the holomorphic representation. Let us consider two anticommuting variables a^* and a,

$$a^* a + a a^* = 0; \qquad (a^*)^2 = 0; \qquad a^2 = 0. \tag{5.3}$$

Such operators are called the generators of the Grassman algebra. The common element of this algebra (a function of the generators) is given by the formula

$$f(a^*, a) = f_{00} + f_{01} a + f_{10} a^* + f_{11} a a^*, \tag{5.4}$$

where $f_{00}, f_{01}, f_{10}, f_{11}$ are complex numbers. We shall call functions depending only on a^* holomorphic functions:

$$f(a^*) = f_0 + f_1 a^*. \tag{5.5}$$

The set of such functions forms a two-dimensional space, and we shall use them for representation of the state vectors of our system.

We shall take the operators a^* and a in the form

$$a^* f(a^*) = a^* f(a^*), \qquad a f(a^*) = \frac{d}{da^*} f(a^*), \tag{5.6}$$

where differentiation is defined naturally by the formula

$$\frac{d}{da^*}(f_0 + f_1 a^*) = f_1. \tag{5.7}$$

It is easy to verify that the commutation relations (5.1) are indeed satisfied. Our next task will be to introduce scalar product in the space of holomorphic functions such that a^* and a will be complex conjugates of each other. We shall do this by means of a convenient definition of the integral of functions of the form (5.4) over $da^* da$. We shall assume that da^* and da anticommute with each other as well as with a and A^*, and we define the following simple integrals:

$$\int a^* da^* = 1; \quad \int a\, da = 1; \quad \int da^* = 0; \quad \int da = 0. \tag{5.8}$$

We may point out that the explicit meaning of the last two formulas consists in the fact that the integral of a total derivative is equal to zero. The said rules are sufficient for defining an integral of any function, if we also stick to the convention that a multiple integral is understood to be a repeated intetral. Then

$$\int f(a^*, a) da^*\, da = f_{11}. \tag{5.9}$$

The scalar product to be found is given by the formula

$$(f_1, f_1) = \int (f_1(a^*))^* f_2(a^*) e^{-a^* a} da^*\, da. \tag{5.10}$$

Here it is to be understood that

$$(f(a^*))^* = f_0^* + f_1^* a. \tag{5.11}$$

Let us check that this scalar product is positive definite. For this purpose, we shall show that the monomials

$$\psi_0 = 1, \psi_1 = a^* \tag{5.12}$$

are orthonormalized. We have

$$(\psi_0, \psi_0) = \int e^{-a^* a}\, da^*\, da = \int (1 - a^* a) da^*\, da = 1, \tag{5.13}$$

$$(\psi_0, \psi_1) = \int a^* e^{-a^* a}\, da^*\, da = 0, \tag{5.14}$$

$$(\psi_1, \psi_1) = \int a a^* e^{-a^* a}\, da^*\, da = 1. \tag{5.15}$$

That the operators a^*, a are conjugates of each other follows from the fact that they are given by the matrices (5.2) in the basis ψ_0, ψ_1. indeed

$$\begin{aligned} a^* \psi_0 &= \psi_1, \qquad & a^* \psi_1 &= 0, \\ a\psi_0 &= 0; \qquad & a\psi_1 &= \psi_0. \end{aligned} \tag{5.16}$$

Let us apply the integration rules formulated above to calculate the integral of an exponential function whose argument is a nonhomogeneous quadratic form:

$$\int \exp\{a^* A a + a^* b + b^* a\} da^*\, da, \tag{5.17}$$

where b and b^* anticommute between themselves and a^*, a.

As follows from (5.9), we can shift the integration variables in the integral (5.17):

$$a^* \to a^* - A^{-1} b^*; \qquad a \to a - A^{-1} b, \tag{5.18}$$

since the coefficient of aa^* in the integrand does not change under such a shift. After this shift the integral (5.17) acquires the form

$$\exp\{-b^* A^{-1} b\} \int \exp\{a^* A a\}\, da^*\, da = -A \exp\{-b^* A^{-1} b\}. \tag{5.19}$$

Note that the formula (5.19) looks exactly as in the case of integration over commuting variables, except that the factor is in the numerator instead of in the denominator as it would be in the case of commuting variables.

We shall now proceed to describe methods of defining operators in the representation under consideration. An operator of the general form may be determined as

$$A = K_{00} + K_{10} a^* + K_{01} a + K_{11} a^* a. \tag{5.20}$$

Two functions can be associated with it on the Grassman algebra: the normal symbol

$$K(a^*, a) = K_{00} + K_{10} a^* + K_{01} a + K_{11} a^* a \tag{5.21}$$

and the kernel

$$A(a^*, a) = A_{00} + A_{10}a^* + A_{01}a + A_{11}a^*a, \tag{5.22}$$

where $A_{nm}, n, m = 0, 1$, are the matrix elements of the operator A in the base ψ_0, ψ_1:

$$A_{nm} = \langle \psi_n | A | \psi_m \rangle. \tag{5.23}$$

It is obvious that

$$(Af)(a^*) = \int A(a^*, \alpha) f(\alpha^*) e^{-\alpha^*\alpha} \, d\alpha^* d\alpha; \tag{5.24}$$

$$(A_1 A_2)(a^*, a) = \int A_1(a^*, \alpha) A_2(\alpha^*, a) e^{-\alpha^*\alpha} d\alpha^* \, d\alpha. \tag{5.25}$$

In order to write these formulas, we have had to introduce new anticommuting variables α^*, α. By definition, α^*, α anticommute with a^*, a.

The normal symbol $K(a^*, a)$ and the kernel $A(a^*, a)$ of the given operator A are related by the formula

$$A(a^*, a) = e^{a^*a} K(a^*, a). \tag{5.26}$$

To prove this statement, it is sufficient to compare the coefficients K_{nm} and A_{nm} in the formulas (5.21) and (5.22) and verify that

$$K_{00} = A_{00}; \quad K_{01} = A_{01}; \quad K_{10} = A_{10}; \quad K_{11} = A_{11} - A_{00} \tag{5.27}$$

All the above formulas are readily generalized to the case of n degrees of freedom. For this purpose one must use $2n$ anticomming variables

$$a_1, \ldots, a_n; \qquad a_1^*, \ldots, a_n^*. \tag{5.28}$$

The space of state vectors consists of analytic functions $f(a^*)$ and has dimension equal to 2^n. The operators $a_i^*, a_i, i = 1, \ldots, n$, act according to the rule

$$a_i f(a^*) = \left(\frac{\partial}{\partial a_i^*} \right)_L f(a^*), \quad a_i^* f(a^*) = a_i^* f(a^*), \tag{5.29}$$

where the subscript L signifies that in differentiating with respect to a_i^* in the function $f(a^*)$ we must displace the variable a_i^* to the left before dropping it.

The operators introduced satisfy the commutatin relations

$$a_i^* a_k + a_k a_i^* = \delta_{ik}, \quad a_i^* a_k^* + a_k^* a_i^* = 0, \quad a_i a_k + a_k a_i = 0 \tag{5.30}$$

and are conjugates of each other with respect to the scalar product

$$(f_1, f_2) = \int (f_1(a^*))^* f_2(a^*) e^{-\Sigma a^* a} \prod da^* da. \tag{5.31}$$

Here the operation* is defined by the formula

$$(C a_{i_1}^* , \ldots a_{i_r}^*)^* = C^* a_{i_r} \ldots a_{i_1}; \tag{5.32}$$

and the integration is performed as before.

The integral of an arbitrary function $f(a^*, a)$ equals

$$\int f(a^*, a) \prod da^* da = f_{1,\ldots,n,n,\ldots,1}\,, \tag{5.33}$$

where $f_{1,\ldots,n,n,\ldots,1}$ is the coefficient of the monomial $a_1 \ldots a_n a_n^* \ldots a_1$ in the expansion of f in the generators. the Gaussian integral

$$\int \exp\{a_i^* A_{ik} a_k + a_i^* b_i + b_i^* a_i\} \prod_i da_i^* da_i \tag{5.34}$$

is calculated by shifting, as in the case of one degree of freedom, and is equal to

$$\exp\{-b_i^* (A^{-1})_{ik} b_k\} \int \exp\{a_i^* A_{ik} a_k\} \prod_i da_i^* da_i\,. \tag{5.35}$$

The remaining integral, in view of (5.33), is equal to $\det A(-1)^n$. Note that the exponential function in the answer may be calculated by substituting into the integrand the solution of the equations

$$\begin{aligned} \left(\frac{d}{da_i^*}\right)_L (a_i^* A_{ik} a_k + a_i^* b_i + b_i^* a_i) &= 0 \\ \left(\frac{d}{da_i}\right)_L (a_i^* A_{ik} a_k + a_i^* b_i + b_i^* a_i) &= 0. \end{aligned} \tag{5.36}$$

This property is general for Gaussian integrals both over usual commuting and over anticommuting variables. We shall frequently use it in the future.

The monomials

$$\psi_{i_1 \ldots i_r} = a_{i_1}^*, \ldots a_{i_r}^* (i_1 < i_2 < \ldots < i_r) \tag{5.37}$$

are orthonormalized and constitute a basis in the space of states.

As in the case of one degree of freedom, an arbitrary operator A can be expressed by the normal symbol $K(a^*, a)$ or by the kernel $A(a^*, a)$. If the operator A is given by the expression

$$A = \sum_{r,t} \sum_{\substack{i_1 < \ldots < i_r \\ j_i < \ldots < j_t}} K_{i_1 \ldots i_r | j_1 \ldots j_t} a_{i_1}^* \ldots a_{i_r}^* a_{j_1} \ldots a_{j_t}\,, \tag{5.38}$$

then

$$K(a^*, a) = \sum_{r,t} \sum_{\substack{i_1 < \ldots < i_r \\ j_i < \ldots < j_t}} K_{i_1 \ldots i_r | j_1 \ldots j_t} a_{i_1}^* \ldots a_{i_r}^* a_{j_1} \ldots a_{j_t} \tag{5.39}$$

and

$$A(a^*, a) = \sum_{r,t} \sum_{\substack{i_1 < \ldots < i_r \\ j_1 < \ldots < j_t}} A_{i_1 \ldots i_r | j_1 \ldots j_t} a_{i_1}^* \ldots a_{i_r}^* a_{j_1} \ldots a_{j_t}\,, \tag{5.40}$$

where

$$A_{i_1 \ldots i_r | j_1 \ldots j_t} = < \psi_{i_1 \ldots i_r} | A | \psi_{j_i \ldots j_t} > . \tag{5.41}$$

The kernel and the normal symbol are related to each other by

$$A(a^*,a) = e^{\Sigma a_i^* a_i} K(a^*,a). \tag{5.42}$$

The action of the operator on a function, and the product of operators, are given by the formulas

$$(Af)(a^*) = \int A(a^*,\alpha) f(\alpha^*) e^{-\Sigma\alpha^*\alpha} \prod d\alpha^* d\alpha, \tag{5.43}$$

$$(A_1 A_2)(a^*,a) = \int A_1(a^*,\alpha) A_2(\alpha^*,a) e^{-\Sigma\alpha^*\alpha} \prod d\alpha^* d\alpha. \tag{5.44}$$

Comparison of the above formulas with the formulas (2.12) and (2.15) derived in Section 2.2 for Bose systems reveals that they have the same form. By following the derivation of the representation for the kernel of the evolution operator in terms of the path integral, one may verify that this derivation is based entirely on the two formulas (2.18) and (2.15). In the case of fermions we have absolutely identical formulas (5.42) and (5.44). Therefore, the representation for the kernel of the evolution operator for a Fermi system with the Hamiltonian $h(a^*,a,t)$ can be immediately written as

$$U(a^*,a;t'',t') = \int \exp\left\{\frac{1}{2}\sum_k (a_k^*(t'')a_k(t'') + a_k^*(t')a_k(t'))\right.$$

$$\left. + i\int_{t'}^{t''} \left[\frac{1}{2i}\sum_k (a_k^*\dot{a}_k - \dot{a}_k^* a_k) - h(a^*(t),a(t),t)\right] dt \prod_{t,k}\right\} da^* da, \tag{5.45}$$

where we assume that

$$a_k^*(t'') = a_k^*, \qquad a_k(t') = a_k. \tag{5.46}$$

We emphasize that we are dealing here with an integral over the infinite-dimensional Grassman algebra with independent generators $a_k^*(t), a_k(t), k = 1,\ldots,n$ for each $t, t' \le t \le t''$.

Let us pass now to the field theory. A complex spinor field may be regarded as a system of fermions with an infinite number of degrees of freedom. In this case the generators of the Grassman algebra are the anticommuting functions $\bar{\psi}(x), \psi(x)$, while the classical action has the form

$$\int \mathcal{L}(x)dx = \int [i\bar{\psi}(\gamma^\mu\partial_\mu + im)\psi - V(\bar{\psi},\psi\varphi)]dx. \tag{5.47}$$

Here γ_μ are the well-known Dirac matrices satisfying the commutation relations

$$[\gamma^\mu,\gamma^\nu]_+ = 2g^{\mu\nu}. \tag{5.48}$$

As the matrices γ one may choose, for example,

$$\gamma_0 = \begin{pmatrix} I & 0 \\ 0 & -I \end{pmatrix}, \quad \gamma^1 = \begin{pmatrix} 0 & I \\ -I & 0 \end{pmatrix}, \quad \gamma^2 = \begin{pmatrix} 0 & \tau_2 \\ -\tau_2 & 0 \end{pmatrix}, \quad \gamma^3 = \begin{pmatrix} 0 & \tau_3 \\ -\tau_3 & 0 \end{pmatrix}, \tag{5.49}$$

where τ_i are the Pauli matrices (formula (1.13) of Chapter 1) and I is the unit matrix. The symbol $\bar{\psi}$ denotes the Dirac adjoint field, $\bar{\psi} = \psi^*\gamma^0$, while the symbol φ stands for all the remaining fields.

In the absence of interaction ($V = 0$) the field ψ satisfies the Dirac equation

$$\left(i\gamma^\mu \frac{\partial}{\partial x^\mu} - m\right)\psi = 0. \tag{5.50}$$

Introduction of the Fourier transform permits writing the Dirac equation in the form

$$\psi(x) = \frac{1}{(2\pi)^{3/2}} \int dK\{e^{i(k_0x_0-kx)}\psi^+(k)+e^{-i(k_0x_0-kx)}\psi^-(k)\}, \quad k_0 = \sqrt{k^2+m^2}, \tag{5.51}$$

where the complex amplitudes $\psi \pm (\vec{k})$ satisfy the equations

$$(k_\mu\gamma_\mu + m)\psi_+(k) = 0, \quad (k_\mu\gamma_\mu - m)\psi_-(k) = 0, \quad k_0 = \sqrt{k^2+m^2}. \tag{5.52}$$

Each of the equations (5.52) possesses two linearly independent solutions for a fixed k. Thus, for instance, in the rest frame, $k = 0$, the equations (5.52) assume the form

$$(\gamma_0 + I)\psi^+(0) = 0, \quad (\gamma_0 - I)\psi^-(0) = 0. \tag{5.53}$$

The solution of these equations, with account taken of the explicit form of the γ-matrices (5.49), depends on two independent constants:

$$\begin{aligned} \psi^+_\alpha(0) &= u_1\delta^3_\alpha + u_2\delta^4_\alpha, \\ \psi^-_\alpha(0) &= \nu_1\delta^1_\alpha + \nu_2\delta^2_\alpha. \end{aligned} \tag{5.54}$$

The solutions in an arbitrary reference frame can be obtained with the aid of the Lorentz transformations. We shall denote the linearly independent solutions of equations (5.52) by $u^i_\alpha(k)$ and $v^i_\alpha(k), i = 1,2$, respectively. The normalization conditions can be chosen so that the relations

$$\begin{aligned} &\bar{u}^i(k)u^j(k) \equiv \sum_\alpha \bar{u}^i_\alpha u^j_\alpha = -\frac{m}{k_0}\delta^{ij}, \quad \bar{u}^i(k)\nu^j(k) = 0, \\ &\bar{v}^i(k)\nu^j(k) = \frac{m}{k_0}\delta^{ij}, \quad \bar{v}^i(k)u^j(k) = 0 \end{aligned} \tag{5.55}$$

are fulfilled. Here $\bar{u} = u^*\gamma^0, \bar{v} = \nu^*\gamma^0$.

It is useful, also, to present the formulas for summation over the spin index:

$$\sum_{i=1,2} u^i(k) \bigotimes \bar{u}^i(k) = \frac{\hat{k} - m}{2k_0}, \tag{5.56}$$

$$\sum_{i=1,2} \nu^i(k) \bigotimes \bar{v}^i(k) = \frac{\hat{k} + m}{2k_0}. \tag{5.57}$$

Since the spinors u and v constitute a complete ortho-normalized set, an arbitrary spinor function $\psi(x)$ can be represented in the form

$$\psi_\alpha(x) = \frac{1}{(2\pi)^{3/2}} \int dk[e^{-ikx} u^i_\alpha(k) b^*_i(k) + e^{ikx} \nu^i_\alpha(k) c_i(k)], \tag{5.58}$$

$$\psi^*_\alpha(x) = \frac{1}{(2\pi)^{3/2}} \int dk[e^{ikx} u^{i*}_\alpha(k) b_i(k) + e^{-ikx} \nu^{i*}_\alpha(k) c^*_i(k)]. \tag{5.59}$$

If $\psi_\alpha(x)$ is a fermion field, then the coefficients $b(k), b^*(k), c(k), c^*(k)$ are to be considered generators of the Grassman algebra, which are numbered according to the values of the momentum k. In terms of the generators b, b^*, c, c^* the normal product is defined in the usual way; that is, in the expressions for arbitrary operators in terms of the kernel $A(b^*, c^*; b, c)$ or of the normal symbol $K(b^*, c^*; b, c)$ the generators b^*, c^* are placed to the left of b, c.

Let us consider a system of Fermi fields interacting with an external source. As an external source we shall take anticommuting spinor functions $\xi(x), \bar{\xi}(x)$. The Hamiltonian of such a system has the form

$$\begin{aligned} h = & \int (i\bar{\psi}(x)\gamma_k \partial_k \psi(x) + m\bar{\psi}(x)\psi(x) + \bar{\psi}(x)\xi(x) + \bar{\xi}(x)\psi(x)) d^3x \\ = & \int [\sqrt{k^2+m^2}(b^*_i(k)b_i(k) + c^*_i(k)c_i(k)) + \gamma^*_i(k,t)b_i(k) + b^*_i(k)\gamma_i(k,t) \\ & + \delta^*_i(k,t)c_i(k) + c^*_i(k)\delta_i(k,t)] d^3k. \end{aligned} \tag{5.60}$$

Here we have used the orthonormality properties of the spinors u_i and v_i, $i = 1, 2$, in passing to the momentum representation. Then

$$\begin{gathered} \gamma_i(k,t) = u^*_i \bar{\xi}(k,t), \quad \delta_i(k,t) = v^*_i \bar{\xi}(k,t), \\ \bar{\xi}(k,t) = \left(\frac{1}{2\pi}\right)^{3/2} \int \xi(x,t) e^{ikx} d^3x. \end{gathered} \tag{5.61}$$

The S-matrix in the form of a path integral is given by the formula

$$\begin{aligned} S(b^*, c^*; b, c) = & \lim_{\substack{t'' \to \infty \\ t' \to -\infty}} \int \exp\Big\{ \frac{1}{2} \int d^3k (b^*_i(k,t'') b_i(k,t'') + b^*_i(k,t') b_i(k,t') \\ & + (b \leftrightarrow c) + i \int_{t'}^{t''} dt \Big[\int d^3k \frac{1}{2i} (\dot{b}^*_i(k,t) b_i(k,t) - b^*_i(k,t) \dot{b}_i(k,t) + (b \leftrightarrow c)) \\ & - h(b^*, b, c^*, c) \Big] \Big\} \prod_{k,t} \frac{db^*(k,t) db(k,t)}{2\pi i} \times \frac{dc^*(k,t) dc(k,t)}{2\pi i}, \end{aligned} \tag{5.62}$$

where

$$\begin{aligned} b^*(k,t'') &= b^*(k) \exp\{i\omega(k)t''\}, \\ b(k,t') &= b(k) \exp\{-i\omega(k)t'\}, \\ e^*(k,t'') &= c^*(k) \exp\{i\omega(k)t''\}, \\ c(k,t') &= c(k) \exp\{-i\omega(k)t'\}. \end{aligned} \tag{5.63}$$

This integral is Gaussian and is calculated exactly in the same way as the corresponding integral (3.11), (3.20) for the scalar field. The expression for the normal symbol is written in the explicitly relativistic form

$$S_{\text{norm}}(\bar{\xi},\xi,b^*,b,c^*,c) = \exp\left\{ i\int \bar{\xi}(x)S_c(x-y)\xi(y)dx\,dy \right.$$
$$\left. +i\int(\bar{\xi}(x)\psi_0(x)+\bar{\psi}_0(x)\xi(x))dx \right\}, \qquad (5.64)$$

where

$$S_c(x-y) = \left(\frac{1}{2\pi}\right)^3 \frac{1}{i}\int [e^{-ik(x-y)}v_i(k)\bigotimes \bar{v}_i(k)\theta(t_2-t_1)$$
$$- e^{ik(x-y)}u_i(k)\bigotimes \bar{u}_i(k)\theta(t_1-t-2)]|_{k_0=\omega}d^3k$$
$$= +\left(\frac{1}{2\pi}\right)^4 \int (\gamma_\mu k_\mu - m + i0)^{-1}e^{-ik(x-y)}d^4k \qquad (5.65)$$

is the causal Green function of the dirac equation and

$$\psi_0(x) = \left(\frac{1}{2\pi}\right)^{3/2} \int (b_i^*(k)u_i(k)e^{ikx} + v_i(k)c_i(k)e^{ikx})i\big|_{k_0=\omega} d^3k, \qquad (5.66)$$

is the solution of the free Dirac equation

$$(i\gamma_\mu\partial_\mu - m)S_c(x) = \delta(x), \qquad (5.67)$$

$$(i\gamma_\mu\partial_\mu - m)\psi_0 = 0. \qquad (5.68)$$

In order to pass from the first representation for S_c to the second one, we again use the formulas (5.56) and (5.57) for summation over the spin index. The first representation makes it obvious in what sense the Green function S_c is causal:

$$\psi_1(x) = \int S_c(x-y)\xi(y)\,d^4y, \qquad (5.69)$$

where ξ decreases rapidly as $|t| \to \infty$, and has no positive (negative) frequency waves as $t \to -\infty\,(t \to +\infty)$.

The formula (5.64) can be taken as the basis for the derivation of the perturbation-theory expansion for the S-matrix of the spinor field, interacting with itself or with other fields. For this note that

$$\left(\frac{1}{i}\frac{\delta}{\delta\bar{\xi}(x)}\right)_L \exp\left\{i\int(\bar{\xi}\psi+\bar{\psi}\xi)dx\right\} = \psi(x)\exp\left\{i\int(\bar{\xi}\psi+\bar{\psi}\xi)dx\right\},$$
$$\left(\frac{1}{i}\frac{\delta}{\delta\xi(x)}\right)_R \exp\left\{i\int(\bar{\xi}\psi+\bar{\psi}\xi)dx\right\} = \exp\left\{i\int(\bar{\xi}\psi+\bar{\psi}\xi)dx\right\}\bar{\psi}(x), \qquad (5.70)$$

where the definition of the right-hand derivative modifies in a natural way the definition of the left-hand one. These formulas together with (5.64) allow one to reduce the path integral for the S-matrix with an arbitrary interaction of the fields $\bar{\psi}, \psi, \varphi$ to the integral for the S-matrix with an external source. The Green functions of the spinor field and the reduction formulas are obtained by a natural modification of the formula for the scalar field. We shall conclude our discussion by giving the expression for the generating functional for the Green functions for interacting spinor and scalar fields with the Lagrangian

$$\mathcal{L}(x) = \bar{\psi}(x) i\gamma_\mu \partial_\mu \psi(x) - m\bar{\psi}(x)\psi(x) + \frac{1}{2}\partial_\mu \varphi(x)\partial_\mu(x)$$
$$-\frac{1}{2}m^2\varphi^2(x) - g\bar{\psi}(x)\psi(x)\varphi(x), \qquad (5.71)$$

containing the most simple version of interaction; this functional is given by the expression

$$Z(\eta, \bar{\xi}, \xi) = \exp\left\{-ig \int \frac{1}{i}\left(\frac{\delta}{\delta\bar{\xi}(x)}\frac{\delta}{\delta\xi(x)}\frac{\delta}{\delta\eta(x)}\right) dx\right\}$$
$$\times \exp\left\{i \int (\bar{\xi}(x) S_c(x-y)\xi(y) + \frac{1}{2}\eta(x) D(x-y)\eta(y)) dx\, dy\right\} \qquad (5.72)$$

and can be written down in the form of the path integral

$$Z(\eta, \bar{\xi}, \xi) = \frac{1}{N}\int \exp\left\{i\int (\mathcal{L}(x) + \eta\varphi + \bar{\xi}\psi + \bar{\psi}\xi) dx\right\} \prod_x d\bar{\psi} d\psi d\varphi, \qquad (5.73)$$

where the integration runs over the fields $\bar{\psi}(x), \psi(x), \varphi(x)$ satisfying the emission conditions.

2.6 The Properties of the Path Integral in Perturbation Theory

As we already mentioned above, at present there is no definition of the path integral in internal terms. However, for the purposes of perturbation theory in quantum field theory it is sufficient to be able to work with path integrals of a special type - namely, with Gaussian integrals. for such integrals it is possible to develop a calculation and transformation technique, which contains in a compact and clear form all the combinatories of the diagram technique of perturbation theory.

Let us obtain these rules in the case of a scalar field, taking for an example the generating functional for the Green functions. For this functional we have two equivalent representations: in the form of a path integral (4.21) and the explicit formula (3.36). We shall use the formula (4.1) as the definition of the Gaussian path integral. More precisely, we shall assume that

$$\int \exp\left\{i\int \frac{1}{2}\varphi(x)K(x-y)\varphi(y)dx\,dy + i\int \varphi(x)\eta(x)dx\right\}$$
$$\times\varphi(x_1)\dots\varphi(x_n)\prod_x d\varphi(x) = (-i)^n \frac{\delta}{\delta\eta(x_1)}$$
$$\dots\frac{\delta}{\delta\eta(x_n)}\exp\left\{-\frac{i}{2}\int \eta(x)K^{-1}(x-y)\eta(y)dx\,dy\right\}, \quad (6.1)$$

where the kernel $K(x-y)$ may, without loss of generality, be considered a symmetric function of its arguments.

By definition, integration over $d\varphi$ is interchangeable with integration over dx and with differentiation with respect to external sources $\eta(x)$. It is assumed that an operator K with a kernel $K(x-y)$ has an inverse K^{-1} with a kernel $K^{-1}(x-y)$:

$$\int K(x-z)K^{-1}(z-y)dx = \int K^{-1}(x-z)K(z-y)dx = \delta(x-y). \quad (6.2)$$

We shall consider the kernel $K^{-1}(x-y)$ to be a sufficiently smooth function. The function $D_c(x-y)$ which enters into the generating functional $Z(\eta)$, does not, of course, have this property; this leads to the appearance of ultraviolet divergences when the functional is calculated. These divergences are removed by the renormalization procedure, which will be discussed in Chapter 4. The first step in this procedure consists of introducing an intermediate regularization, which substitutes the function $D_c(x-y)$ by a smooth function. Hence, the reasoning that now follows has to do with the regularized perturbation theory.

The class of functions $\varphi(x)$ over which integration is performed must provide a unique definition of the operator inverse to K. If $K = \Box$, then such a condition is the already mentioned causality condition: φ behaves asymptotically for $|t| \to \infty$ as the solution of the free equation which has no incoming waves as $t \to -\infty$ and no outgoing waves as $t \to \infty$. The operator K^{-1} in this case is integral with the kernel D_c (more precisely, as just mentioned, with its regularization). We shall call the integrand in this formula (6.1) at $\eta = 0$ a Gaussian functional.

We shall now pass to the discussion of the path integral defined by the formula (6.1). First, note that the functional (6.1), which it is natural to call the Fourier transform of the Gaussian functional

$$F(\varphi) = \exp\left\{i\int \frac{1}{2}\varphi(x)K(x-y)\varphi(y)dx\,dy\right\}\varphi(x_1)\dots\varphi(x_n), \quad (6.3)$$

is itself a Gaussian functional, since by differentiating $\exp\{\int \eta K^{-1}\eta dx\,dy\}$, we obtain an expression in which this exponential function is multiplied by a polynomial.

We shall show that our definition allows one to prove for the integral (6.1) the validity of the simplest transformations such as integration by parts and change of variables, and we shall also introduce the concept of the functional δ-function.

1. *Integration by parts.* Let us consider the integral

$$I = \int \left[\frac{\delta}{\delta\varphi(z)} \exp\left\{\frac{i}{2}\int \varphi(x)K(x-y)\varphi(y)dx\,dy\right\}\right]$$
$$\times \exp\left\{i\int \varphi(x)\eta(x)dx\right\}\prod_x d\varphi. \tag{6.4}$$

The functional

$$\frac{\delta}{\delta\varphi(z)} \exp\left\{\frac{i}{2}\int \varphi(x)K(x-y)\varphi(y)dx\,dy\right\} \tag{6.5}$$

is Gaussian; therefore the integral (6.4) makes sense and by definition is equal to

$$I = i\int \left[\int K(z-y)\varphi(y)dy\right.$$
$$\left.\times \exp\left\{\frac{i}{2}\int \varphi(x)K(x-y)\varphi(y)dx\,dy + i\int \varphi(x)\eta(x)dx\right\}\right] \times \prod_x d\varphi$$
$$= \int K(z-y)\frac{\delta}{\delta\eta(y)}Z(\eta)dy = -i\eta(z)Z(\eta). \tag{6.6}$$

On the other hand,

$$-i\eta(z)Z(\eta) = -\int \exp\left\{\frac{i}{2}\int \varphi(x)K(x-y)\varphi(y)dx\,dy\right\}$$
$$\times \frac{\delta}{\delta\varphi(z)}\exp\left\{i\int \varphi(x)\eta(x)dx\right\}\prod_x d\varphi \tag{6.7}$$

Comparing (6.4) and (6.6), we see that we have the usual formula for integration by parts, and the boundary terms are dropped. This result is generalized in an obvious way to an arbitrary Gaussian integral, since any such integral may be represented as the derivative of I with respect to η.

2. *Repeated integrals.* Since an integral of a Gaussian functional is itself a Gaussian functional, it is possible to define a repeated integral. Let us show that

$$\int \exp\left\{i\sum_{i,j=i}^{n}\frac{(K_n)_{ij}^{xy}}{2}\varphi_i^x\varphi_j^y + \sum_{j=1}^{n}\eta_j^x\varphi_j^x\right\}\prod_x d\varphi_1\ldots d\varphi_n$$
$$= \exp\left\{-\frac{i}{2}\sum_{i,j=1}^{n}(K_n^{-1})_{ij}^{xy}\eta_i^x\eta_j^y\right\} \tag{6.8}$$

(we use here abbreviated notation, having in mind that integration is performed over continuous indices x, y).

Let the equality (6.8) be valid for some n. We shall prove that it is valid also for the number $n+1$. According to our assumption

$$I_{n+1} = \int \exp\left\{-i \sum_{i,j=1}^{n} (\eta_i^x + K_{i\,n+1}^{xy}\varphi_{n+1}^y) \times \frac{(K_n^{-1})_{ij}^{xz}}{2} (\eta_j^z + K_{jn+1}^{zu}\varphi_{n+1}^u) + \frac{i}{2} K_{n+1\,n+1}^{xy}\varphi_{n+1}^x\varphi_{n+1}^y + i\eta_{n+1}^x\varphi_{n+1}^x\right\} \prod_x d\varphi_{n+1}. \tag{6.9}$$

Integrating over φ_{n+1}, we obtain

$$I_{n+1} = \exp\left\{-\frac{i}{2}\left(\eta_{n+1}^x - \sum_{i,j=1}^{n} K_{i\,n+1}^{xy}(K_n^{-1})_{ij}^{yz}\eta_j^z\right) \times \left(K_{n+1\,n+1}^{zu} - \sum_{l,m=1}^{n} K_{l\,n+1}^{zv}(K_n^{-1})_{lm}^{vs}K_{m\,n+1}^{su}\right)^{-1} \times \left(\eta_{n+1}^u - \sum_{i,j=1}^{n} K_{i\,n+1}^{ur}(K_n^{-1})_{ij}^{rt}\eta_j^t\right) - \frac{i}{2}\sum_{i,j=1}^{n} \eta_i^x (K_n^{-1})_{ij}^{xy}\eta_j^y\right\}, \tag{6.10}$$

Taking advantage of

$$(K^{-1})_{ij}^{xy} = (\det K^{-1})^{xz}\bar{K}_{ij}^{zy}, \tag{6.11}$$

where K_{ij} stands for the adjoint of the K_{ij}th element of the matrix K, one can represent the second factor in the power of the exponential function as

$$(\det K_n)^{zx}[(\det K_n)^{xy} K_{N=1\,n+1}^{yu} - \sum_{i,j=1}^{n} K_{i\,n+1}^{xy}\bar{K}_{ij}^{uv}K_{j\,n+1}^{vu}]^{-1} = (\det K_n)^{zx}(\det K_{n+1}^{-1}. \tag{6.12}$$

Let us consider the separate terms in the formula (6.10):

$$\eta_{n+1}^x\eta_{n+1}^y(\det K_n)^{xz}(\det K_{n+1}^{-1})^{zy} = \eta_{n+1}^x(K_{n+1}^{-1})_{n+1\,n+1}^{xy}\eta_{n+1}^y,$$

$$\sum_i \eta_{n+1}^x(\det K_{n+1}^{-1})^{xy}K_{i\,n+1}^{yz}\bar{K}_{ij}^{zu}\eta_j^u = \eta_{n_1}^x(\det K_{n+1}^{-1})^{xy}\bar{K}_{n+1\,j}^{yz}\eta_j^z$$

$$= \eta_{n+1}^x(K_{n+1}^{-1})_{n+1\,j}^{xy}\eta_j^y. \tag{6.13}$$

In an analogous manner it is possible to show that the coefficients of $\eta_i\eta_j$ are $(K_{n+1}^{-1})_{ij}^{xy}$. As a result, we obtain

$$I_{n+1} = \exp\left\{-\frac{i}{2}\sum_{i,j=1} (K_{n+1}^{-1})_{ij}^{xy}\eta_i^x\eta_j^y\right\}, \tag{6.14}$$

which was to be proved. Obviously, the result does not depend on the order of integration, since a change in the order is just equivalent to the rearrangement of the columns of the matrix K. Therefore, it is proved that the repeated integrals exist, and the result does not depend on the order of integration.

3. *Definition of the δ-function.*

$$\int \exp\left\{i\int \eta(x)\left[\int c(x-y)\varphi(y)dy - \varphi'(x)\right]dx\right\}\prod_x d\eta$$
$$\overset{\text{def}}{=} \delta\left(\varphi(x) - \int c^{-1}(x-y)\varphi'(y)dy\right). \quad (6.15)$$

This equality means that

$$\int F(\varphi)\left[\int \exp\left\{i\int \eta(x)\left[\int c(x-y)\varphi(y)dy - \varphi'(x)\right]dx\right\} \times \prod_x d\eta\right]\prod_x d\varphi$$
$$\overset{\text{def}}{=} \int\left[\int F(\varphi)\exp\left\{i\int \eta(x)\left[\int c(x-y)\varphi(y)\right.\right.\right.$$
$$\left.\left.\left.-\varphi'(x)\right]dx\right\}\prod_x d\varphi\right]\prod_x d\eta = F(c^{-1}\varphi'), \quad (6.16)$$

where $F(\varphi)$ is a Gaussian functional. By analogy to the usual definition of the δ-function, one might expect that in (6.16) there should be still another constant (that is, independent of φ') factor $\det c^{-1}$. The absence of this factor is explained by the fact that our definition of a functional integral (6.1) includes the normalization condition.

The equality (6.16) is verified by direct calculation:

$$\int\left[\int \exp\left\{\frac{i}{2}\int \varphi(x)K(x-y)\varphi(y)dx\,dy\right\}\right.$$
$$\left.\times \exp\left\{i\int \eta(x)\left[\int c(x-y)\varphi(y)dy - \varphi'(x)\right]dx\right\} \times \prod_x d\varphi\right]\prod_x d\eta$$
$$= \exp\left\{\frac{i}{2}\int \varphi'(x)c^{-1}K(x-y)c^{-1}\varphi'(y)dx\,dy\right\}. \quad (6.17)$$

We shall show now that

$$\int \exp\left\{i\int [f_x(\varphi) - \varphi'(x)]\eta(x)dx\right\}\prod_x d\eta = \delta(f_x(\varphi) - \varphi'(x)), \quad (6.18)$$

$f_x(\varphi)$ as a function of x belongs to the same class as $\varphi(x)$. The function $f_x(\varphi)$ can be expanded in a formal series of the form

$$f_x(\varphi) = c_0(x) + \varphi(x) + \tilde{f}(\varphi),$$

$$\tilde{f}(\varphi) = g\int c_1(x,y)\varphi(y)\,dy + g^2\int c_2(x,y,z)\varphi(y)\varphi(z)dy\,dz + \ldots \quad (6.19)$$

(For simplicity we assume the coefficient of the first power of φ to be equal to 1. This reasoning is trivially generalized to the case of $c \neq 1$ with the help of the previous formula.) The equation

$$c_0(x) + \varphi(x) + \bar{f}(\varphi) - \varphi'(x) = 0 \tag{6.20}$$

has a unique solution, which may be represented as a formal series in g.

Equation (6.18) means that

$$\int\left[F(\varphi)\exp\left\{i\int[f_x(\varphi) - \varphi'(x)]\eta(x)dx\right\}\det\left\{1+\frac{\delta\bar{f}}{\delta\varphi}\right\}\prod_x d\varphi\right]d\eta = F(\bar{\varphi}), \tag{6.21}$$

where $\bar{\varphi}(\varphi')$ is the solution of the equation (6.20); $\det(1 + \delta\bar{f}/\delta\varphi)$ is, by definition,

$$\det\left\{1+\frac{\delta\bar{f}}{\delta\varphi}\right\} \stackrel{\text{def}}{=} \exp\left\{\operatorname{Tr}\ln\left[1+\frac{\delta\bar{f}(x)}{\delta\varphi(y)}\right]\right\}$$

$$= \exp\left\{\int\left.\frac{\delta\bar{f}(x)}{\delta\varphi(y)}\right|_{x=y}dx - \frac{1}{2}\int dx\,dy\frac{\delta\bar{f}(x)\delta\bar{f}(y)}{\delta\varphi(y)\delta\varphi(x)} + \dots\right\} \tag{6.22}$$

According to the definition,

$$\int\left[\int\exp\left\{\frac{i}{2}\int\varphi(x)K(x-y)\varphi(y)dx\,dy\right.\right.$$

$$\left.\left.+i\int[f(\varphi)-\varphi'(x)]\eta(x)dx\right\}\det\left\{1+\frac{\delta\bar{f}}{\delta\varphi}\right\}\prod_x d\varphi\right]\prod_x d\eta$$

$$= \int\left[\overrightarrow{\exp}\left\{i\int\eta(x)\bar{f}\left(\frac{1}{i}\frac{\delta}{\delta\eta}\right)dx\right\}\det\left\{1+\frac{\delta\bar{f}}{\delta\varphi}\left(\frac{1}{i}\frac{\delta}{\delta\eta}\right)\right\}\right.$$

$$\left.\times\exp\left\{-\frac{i}{2}\int\eta(x)K^{-1}(x-y)\eta(y)dx\,dy\right\}\right]$$

$$\times\exp\left\{-i\int[\varphi'(x)-c_0(x)]\eta(x)dx\right\}\prod_x d\eta. \tag{6.23}$$

The sign $\rightarrow$ ($\leftarrow$) over the exponential function shows that in the representation of the exponential function as a series, all operators $\delta/\delta\eta$ must be placed to the right (left) of η. Ingrating by parts, we transform the right-hand side into

$$\int\exp\left\{-\frac{i}{2}\int\eta(x)K^{-1}(x-y)\eta(y)dx\,dy\right\}$$

$$\times\det\left\{1+\frac{\delta\bar{f}}{\delta\varphi}\left(\frac{1}{i}\frac{\delta}{\delta\varphi}\right)\right\}\overleftarrow{\exp}\left\{i\int\bar{f}\left(-\frac{1}{i}\frac{\delta}{\delta\eta}\right)\eta(x)dx\right\}$$

$$\times\exp\left\{-i\int[\varphi'(x)-c_0(x)]\eta(x)dx\right\}\prod_x d\eta. \tag{6.24}$$

Let us consider the functional

$$B(\varphi',\eta)=\det\left\{1+\frac{\delta\tilde f}{\delta\varphi}\left(\frac{1}{i}\frac{\delta}{\delta\eta}\right)\right\}\overleftarrow{\exp}\left\{i\int\tilde f\left(-\frac{1}{i}\frac{\delta}{\delta\varphi}\right)\eta(x)dx\right\}$$
$$\times\exp\left\{-i\int[\varphi'(x)-c_0(x)]\eta(x)dx\right\}=\overleftarrow{\exp}\left\{-i\int\frac{\delta}{\delta\varphi'(x)}\tilde f(\varphi'-c_0)dx\right\}$$
$$\times\det\left\{1+\frac{\delta\tilde f}{\delta\varphi}(\varphi'-c_0)\right\}\exp\left\{-i\int[\varphi'(x)-c_0(x)]\eta(x)dx\right\}\qquad(6.25)$$

$B(\varphi',\eta)$ satisfies the equation

$$\delta B/\delta\eta(x)=i\left[c_0(x)-\varphi'(x)+\tilde f\left(i\frac{\delta}{\delta\eta}\right)\right]B\qquad(6.26)$$

with the initial condition

$$B(\varphi',0)=A(\varphi')=\exp\left\{-\int\frac{\delta}{\delta\varphi'(x)}\tilde f(\varphi'-c_0)dx\right\}$$
$$\times\det\left\{1+\frac{\delta\tilde f}{\delta\varphi}(\varphi'-c_0)\right\}\cdot 1.\qquad(6.27)$$

We shall look for the solution of the equation (6.26) in the form

$$B(\varphi',\eta)=A(\varphi')\exp\left\{-i\int\varphi(\varphi')\eta(x)\,dx\right\}.\qquad(6.28)$$

Substituting (6.28) into (6.26), we obtain

$$\varphi(x)=\varphi'(x)-c_0(x)-\tilde f(\varphi).\qquad(6.29)$$

Therefore the integral of interest is equal to

$$A(\varphi')\int\exp\left\{-\frac{i}{2}\int\eta(x)K^{-1}(x-y)\eta(y)dx\,dy\right\}$$
$$\times\exp\left\{-i\int\bar\varphi(x)\eta(x)dx\right\}\prod_x d\eta$$
$$=A(\varphi')\exp\left\{\frac{i}{2}\int\bar\varphi(x)K(x-y)\bar\varphi(y)dx\,dy\right\}.\qquad(6.30)$$

It remains only for us to prove that $A(\varphi')=1$. The formula (6.27) can be written as

$$A(\varphi')=\det\left[\overleftarrow{\exp}\left\{-\frac{\delta}{\delta\varphi'(x)}\tilde f(\varphi'-c_0)\right\}\times\left\{1+\frac{\delta\tilde f}{\delta\varphi}(\varphi'-c_0)\right\}\right]\cdot 1$$
$$=\det\left[\sum_n\frac{(-1)^n}{n!}\frac{\delta^n}{\delta\varphi'^n(x)}\tilde f^n(\varphi'-c_0)\left\{1+\frac{\delta\tilde f}{\delta\varphi'}(\varphi'-c_0)\right\}\right]\cdot 1.\qquad(6.31)$$

Let us consider the n-th term of the sum in square brackets. It is a binomial, the first term of which may be represented as

$$\frac{(-1)^n}{n!}\frac{\delta^n}{\delta\varphi'^n}\bar{f}^n(\varphi' = \frac{(-1)^n}{(n-1)!}\frac{\delta^{n-1}}{\delta\varphi'^{n-1}}\left(\frac{\delta\bar{f}}{\delta\varphi'}\bar{f}^{n-1}(\varphi')\right). \tag{6.32}$$

On the other hand, the $(n-1)$th term of the sum in square brackets is represented by an analogous binomial, the second term of which is equal to

$$\frac{(-1)^{n-1}}{(n-1)!}\frac{\delta^{n-1}}{\delta\varphi'^{n-1}}\left(\frac{\delta\bar{f}}{\delta\varphi'}\bar{f}^{n-1}(\varphi')\right). \tag{6.33}$$

Thus, successive terms in square brackets cancel out, and the whole expression is equal to one.

4. *Change of variables.* Let

$$I = \int \exp\left\{\frac{i}{2}\int \varphi(x)K(x-y)\varphi(y)dx\,dy + i\int \varphi(x)\eta(x)dx\right\}\prod_x d\varphi. \tag{6.34}$$

By a change of variables

$$\varphi = f_x(\varphi'), \quad f_x(\varphi') = c_0(x) + \varphi'(x) + \bar{f}(\varphi'), \tag{6.35}$$

I is reduced to the form

$$I = \int \exp\left\{\frac{i}{2}\int f_x(\varphi')K(x-y)f_y(\varphi'dx\,dy\right.$$

$$\left. + i\int f_x(\varphi')\eta(x)dx\right\}\det\left\{1+\frac{\delta\bar{f}}{\delta\varphi'}\right\}\prod_x d\varphi'. \tag{6.36}$$

In order to prove this statement, it is sufficient to verify that the Fourier transforms of (6.34) and of (6.36) are equal to each other. The Fourier transform of (6.34) is

$$\exp\left\{\frac{i}{2}\int \psi(x)K(x-y)\psi(y)dx\,dy\right\}. \tag{6.37}$$

The Fourier transform of (6.36) equals

$$I(\psi) = \int I(\eta)\exp\left\{-i\int \eta(x)\psi(x)dx\right\}\prod_x d\eta$$

$$= \exp\left\{\frac{i}{2}\int f_x K(x-y)f_y dx\,dy\right\}\det\left\{1+\frac{\delta\bar{f}}{\delta\varphi'}\right\} \times \delta(\psi - f(\varphi'))\prod_x d\varphi'$$

$$= \exp\left\{\frac{i}{2}\int \psi(x)K(x-y)\psi(y)dx\,dy\right\}. \tag{6.38}$$

The statement is proved.

Our reasoning shows that all those properties of the Feynman integral which are used in practice in perturbation theory follow directly from the definition of the quasi-Gaussian integral and can be rigorously proved independently of the question of the existence of the Feynman-integral measure. Thus, in the framework of perturbation theory, the formalism of the path integral is a perfectly rigorous mathematical method, and results obtained with it do not need additional justification.

All these conclusions apply to the same extent to path integrals containing Fermi variables. In this case one must bear in mind the anticommutativity of variational derivatives, and in formulas for changing variables the corresponding determinant must be written in the denominator instead of the numerator. This characteristic feature of Gaussian integrals over Fermi variables has been already discussed above.

3

Quantization of the Yang-Mills Field

3.1 The Lagrangian of the Yang-Mills Field and the Specific Properties of its Quantization

In the previous chapter we used the path integral to formulate quantization rules for those scalar and spinor fields chosen as examples. At first sight it seems possible to quantize the Yang-Mills field in an analogous manner, considering each component of the field to be a scalar field. This, however, is not so. Gauge invariance introduces certain specific features into the quantization procedure. The spinor and scalar fields with which the Yang-Mills field interacts do not have any influence on these specific features. In the first three following sections we shall therefore restrict ourselves to the discussion of the Yang-Mills field in a vacuum.

We recall the notatin introduced in the first chapter. Let Ω be a compact group of internal symmetry, $T^a (a = 1, \ldots, n)$ its orthonormalized generators in the adjoint representation, t^{abc} corresponding structure constants, and

$$\mathcal{A}_\mu = A_\mu^a T^a \tag{1.1}$$

the Yang-Mills field. The gauge transformation is given by the matrix $\omega(x)$ with values in the adjoint representation of the group

$$\mathcal{A}_\mu(x) \rightarrow \mathcal{A}_\mu^\omega(x) = \omega(x)\mathcal{A}_\mu(x)\omega^{-1}(x) + \partial_\mu\omega(x)\omega^{-1}(x) \tag{1.2}$$

The gauge-invariant Lagrangian has the form

$$\mathcal{L} = \frac{1}{8g^2} \operatorname{tr}\{\mathcal{F}_{\mu\nu}\mathcal{F}_{\mu\nu}\}, \tag{1.3}$$

where

$$\mathcal{F}_{\mu\nu} = \partial_\nu \mathcal{A}_\mu - \partial_\mu \mathcal{A}_\nu + [\mathcal{A}_\mu, \mathcal{A}_\nu]. \tag{1.4}$$

The equations of motion

$$\nabla_\mu \mathcal{F}_{\mu\nu} = \partial_\mu \mathcal{F}_{\mu\nu} - [\mathcal{A}_\mu, \mathcal{F}_{\mu\nu}] = 0 \tag{1.5}$$

are second-order equations with respect to $\mathcal{A}_\mu$ and are gauge-invariant: If $\mathcal{A}_\mu(x)$ is a solution of the equations of motion, then $\mathcal{A}_\mu^\omega(x)$ will also be a solution for any $\omega(x)$, where ω depends arbitrarily on x. This means that the equations of motion (1.5) are not independent. And, indeed, it is not difficult to verify that

$$\nabla_\nu \nabla_\mu \mathcal{F}_{\mu\nu} = 0 \tag{1.6}$$

To prove this, we represent $\nabla_\nu \nabla_\mu$ as

$$\nabla_\nu \nabla_\mu = \frac{1}{2}(\nabla_\nu \nabla_\mu + \nabla_\mu \nabla_\nu) + \frac{1}{2}(\nabla_\nu \nabla_\mu - \nabla_\mu \nabla_\nu). \tag{1.7}$$

We recall that for any matrix $\mathcal{B}(x)$ in the adjoint representation,

$$(\nabla_\nu \nabla_\mu - \nabla_\mu \nabla_\nu)\mathcal{B}(x) = [\mathcal{F}_{\mu\nu}, \mathcal{B}]. \tag{1.8}$$

Since $\mathcal{F}_{\mu\nu}$ is antisymmetric with respect to μ and ν, we get

$$\nabla_\nu \nabla_\mu \mathcal{F}_{\mu\nu} = \frac{1}{2}[\mathcal{F}_{\mu\nu}, \mathcal{F}_{\mu\nu}] = 0. \tag{1.9}$$

The equality (1.9) is a special case of the second Noether theorem, which asserts that invariance of the Lagrangian with respect to transformations depending on an arbitrary function leads to the linear dependence of the equations of motion.

The above mentioned specific feature of the equations of motion manifests itself in their quantization. Indeed, some of the functions that parametrize the classical solution depend arbitrarily on time and do not obey the dynamics. While quantizing, we must separate the true dynamical variables and the group parameters. We shall deal with this problem in the next section.

We shall now show why we cannot naïvely transfer the rules developed in the previous chapter for the construction of perturbation theory to the case of the Yang-Mills fields. According to the prescription of Chapter 2, in order to construct the perturbation theory for a given Lagrangian $\mathcal{L}$, one must represent it in the form

$$\mathcal{L} = \mathcal{L}_0 + \mathcal{L}_{\text{int}}, \tag{1.10}$$

where $\mathcal{L}_0$ is a form quadratic in the fields, and $\mathcal{L}_{\text{int}}$ contains higher-order forms in th fields. The monomials in $\mathcal{L}_{\text{int}}$ define the vertices with three or more legs, and $\mathcal{L}_0$ defines the propagators corresponding to internal lines. Namely, the propagator is the kernel of the integral operator that is inverse to the differential operator which defines the quadratic form $\mathcal{L}_0$.

Up to an inessential divergence, $\mathcal{L}_0$ for the Yang-Mills Lagrangian has the form (for the normalization $\mathcal{A}_\mu \to g\mathcal{A}_\mu$)

$$\mathcal{L}_0 = -\frac{1}{4}(\partial_\nu A^a_\mu - \partial_\mu A^a_\nu)(\partial_\nu A^a_\mu - \partial_\mu A^a_\nu) = -\frac{1}{2}[\partial_\mu A^a_\nu \partial_\mu A^a_\nu - \partial_\mu A^a_\mu \partial_\nu A^a_\nu]. \quad (1.11)$$

In the momentum representation, the quadratic form $\mathcal{L}_0$ is given by the expression

$$K^{ab}_{\mu\nu}(k) = \delta^{ab}(g_{\mu\nu}k^2 - k_\mu k_\nu). \quad (1.12)$$

This operator has no inverse, and therefore the propagator is not defined. The reason is that, as has already been pointed out, not all components of the Yang-Mills field are independent dynamical variables. An analogous difficulty, as is known, is met in quantum electrodynamics. In this case one uses the Gupta-Bleuler formalism, which actually reduces to the following: The function

$$D_{\mu\nu}(k^2) = -\frac{g_{\mu\nu}}{k^2 + i0} \quad (1.13)$$

is chosen as a photon propagator, and it is shown that the S-matrix constructed with its aid is unitary.

Generalizing this recipe, one could try to construct the Yang-Mills theory, using the propagator

$$D^{ab}_{\mu\nu} = -\frac{\delta^{ab} g_{\mu\nu}}{k^2 + i0}. \quad (1.14)$$

However, as was first shown by Feynman, such a construction of the perturbation theory is inadmissible. The S-matrix calculated with such a propagator is not unitary. Therefore it is necessary to revise the derivation of the perturbation-theory rules proceeding from the causal description of the classical dynamics for the Yang-Mills field, using for it the most convenient Hamiltonian formulation of this theory.

3.2 The Hamiltonian Formulation of the Yang-Mills Field and its Quantization

In order to construct a consistent quantization procedure we must first find the true dynamical variables for the Yang-Mills field and verify that they change with time according to the laws of Hamiltonian dynamics. After this, we shall be able, in constructing the evolution operator, to use the path-integral formalism developed in the previous chapter.

Let us consider in greater detail the structure of the Lagrangian in the first-order formalism:

$$\mathcal{L} = \frac{1}{4}\operatorname{tr}\left\{\left(\partial_\nu \mathcal{A}_\mu - \partial_\mu \mathcal{A}_\nu + g[\mathcal{A}_\mu, \mathcal{A}_\nu] - \frac{1}{2}\mathcal{F}_{\mu\nu}\right)\mathcal{F}_{\mu\nu}\right\}, \quad (2.1)$$

where $\mathcal{A}_\mu$ and $\mathcal{F}_{\mu\nu}$ are assumed to be independent variables. This Lagrangian, and the equations of motion following from it are equivalent to the Lagrangian (1.3). Indeed, the equations of motion for $\mathcal{F}_{\mu\nu}$ following from (2.1) contain no derivatives. Substituting their solution into (2.1), we obtain the Lagrangian (1.3).

In the three-dimensional notation ($\mu = 0, k$; $\nu = 0, l$; $k, l = 1,2,3$) we may rewrite the Lagrangian (up to a divergence) in the form

$$\mathcal{L} = -\frac{1}{2}\text{tr}\left\{\mathcal{E}_k\partial_0\mathcal{A}_k - \frac{1}{2}(\mathcal{E}_k^2 + \mathcal{G}_k^2) + \mathcal{A}_0\mathcal{C}\right\}, \tag{2.2}$$

where

$$\mathcal{E}_k = \mathcal{F}_{k0}, \mathcal{G}_k = \frac{1}{2}\epsilon^{ijk}\mathcal{F}_{ji}, \mathcal{C} = \partial_k\mathcal{E}_k - g[\mathcal{A}_k, \mathcal{E}_k] \tag{2.3}$$

and we assume that $\mathcal{F}_{ik}$ is expressed in terms of $\mathcal{A}_i$ through the equations of motion, not including time derivatives:

$$\mathcal{F}_{ik} = \partial_k\mathcal{A}_i - \partial_i\mathcal{A}_k + g[\mathcal{A}_i, \mathcal{A}_k]. \tag{2.4}$$

This same Lagrangian can be written in the form

$$\mathcal{L} = E_k^a\partial_0A_k^a - h(E_k, A_k) + A_0^aC^a, \quad h = \frac{1}{2}\{(E_k^a)^2 + (G_k^a)^2\}. \tag{2.5}$$

It is clear from the form of the Lagrangian (2.5) that the pairs (E_k^a, A_k^a) are canonical variables; h is the Hamiltonian, A_0^a is the Lagrangian multiplier, and C^a is the constraint on the canonical variables. By introducing Poisson bracets

$$\{E_k^a(x), A_i^b(y)\} = \delta_{ki}\delta^{ab}\delta(x-y), \tag{2.6}$$

it is easy to verify that

$$\{C^a(x), C^b(y)\} = gt^{abc}C^c(x)\delta(x-y) \tag{2.7}$$

and that

$$\left\{\int d^3x[(E_k^a)^2 + (G_k^a)^2], C^b(y)\right\} = 0. \tag{2.8}$$

This means that our system presents an example of the so called generalized Hamilton dynamics. This concept was introduced by Dirac. Let us consider it, using as an example a system Γ with n degrees of freedom. Let p_i and q_i be its canonical variables, running through the phase space Γ^{2n}, and let the action have the form

$$A = \int \left[\sum_{i=1}^{n} p_i\dot{q}_i - h(p,q) - \sum \lambda^\alpha\varphi^\alpha(p,q)\right] dt; \qquad \alpha = 1\ldots m, m < n. \tag{2.9}$$

Here the variables λ^α, addition to p and q, are called the Lagrangian multipliers; and φ^α are the constraints. Such an action defines a generalized Hamiltonian system if the conditions

$$\{h, \varphi^\alpha\} = c^{\alpha\beta}(p,q)\varphi^\beta; \{\varphi^\alpha, \varphi^\beta\} = \sum_\gamma c^{\alpha\beta\gamma}(p,q)\varphi^\gamma \tag{2.10}$$

are fulfilled with certain coefficients $c^{\alpha\beta}$ and $c^{\alpha\beta\gamma}$, which in general depend on p, q. We make use of the term "generalized Hamiltonian system" in a narrow sense, imposing on the constraints the condition (2.10) in accordance with which their Poisson brackets vanish on the surface of constraints. In the general case such a condition may not hold. Constraints such as (2.10) are conventionally called constraints of the first class. Imposition of constraints of the first class is sufficient for our book.

The generalized Hamiltonian system is equivalent to the usual Hamiltonian system Γ^* with $n-m$ degrees of freedom. The phase space $\Gamma^{*(n-m)}$ of the latter system may be realized in the following manner. Let us consider m subsidiary conditions,

$$\chi^m(p, q) = 0, \tag{2.11}$$

for which the requirments

$$\det|\{\varphi^\alpha, \chi^\beta\}| \neq 0, \tag{2.12}$$

$$\{\chi^\alpha, \chi^\beta\} = 0. \tag{2.13}$$

are satisfied. (The condition (2.13) is not essential, but it is convenient; and we shall apply it below.)

The subspace in Γ^{2n}

$$\chi^\alpha(p, q) = 0, \qquad \varphi^\alpha(p, q) = 0 \tag{2.14}$$

is the space $\Gamma^{*(n-m)}$ in question. The canonical variables p^*, q^* in $\Gamma^{*2(n-m)}$ may be found as follows. Owing to the condition (2.13), we can choose the canonical variables in Γ^{2n} in such a way that the x^α will coincide with the first m variables of the coordinate type

$$q = (\chi^\alpha, q^*). \tag{2.15}$$

Let

$$p = (p^\alpha, p^*) \tag{2.16}$$

be the corresponding conjugate momenta. In these variables the condition (2.12) takes the form

$$\det\left|\frac{\partial \Phi^\alpha}{\partial p^\beta}\right| \neq 0, \tag{2.17}$$

so that the equations of the constraints

$$\varphi^\alpha(p, q) = 0 \tag{2.18}$$

may be solved for p^α. As a result, the subspace $\Gamma^{*2(n-m)}$ is given by the equations

$$\chi^\alpha \equiv q^\alpha = 0; \qquad p^\alpha = p^\alpha(p^*, q^*) \tag{2.19}$$

and p^*, q^* are canonical. The Hamiltonian of this system is the function

$$h^*(p^*, q^*) = h(p, q)|_{\varphi=0, \chi=0}. \tag{2.20}$$

The equivalence of the systems Γ and Γ^* means the following. Let us consider the equations of motion for the system Γ:

$$\dot{p}_i = \frac{\partial h}{\partial q_i} + \lambda^\alpha \frac{\partial \varphi^\alpha}{\partial q_i} = 0, \qquad \zeta_i - \frac{\partial h}{\partial p_i} - \lambda^\alpha \frac{\partial \varphi^\alpha}{\partial p_i} = 0, \tag{2.21}$$

$$\varphi^\alpha = 0.$$

The solutions of these equations contain arbitrary functions $\lambda^\alpha(t)$. Subsidiary conditions $\chi^\alpha(p,q) = 0$ remove this arbitrariness, by expressing $\lambda^\alpha(t)$ in terms of the canonical variables. As a result, the equations for the variables p^*, q^* are the only equations left. These equations coincide with the Hamiltonian equations for the system Γ^*,

$$\zeta^* = \frac{\partial h^*}{\partial p^*}; \qquad \dot{p}^* = -\frac{\partial h^*}{\partial q^*}. \tag{2.22}$$

Indeed, consider the equations (2.19), (2.21) in terms of the coordinates (2.15), (2.16). The equations $q^\alpha = 0$ lead to relationships that allow one to find λ^α:

$$\frac{\partial h}{\partial p_\alpha} + \lambda^\beta \frac{\partial \varphi_\beta}{\partial p_\alpha} = 0. \tag{2.23}$$

Let us now consider some one of the coordinates q^* and compare the equations for it, which follow from (2.19), (2.21), and (2.22). They have the form

$$\zeta^* = \frac{\partial h}{\partial p^*} + \lambda^\alpha \frac{\partial \varphi_\alpha}{\partial p^*}, \tag{2.24}$$

$$\zeta^* = \frac{\partial h^*}{\partial p^*} = \frac{\partial h}{\partial p^*} + \frac{\partial h}{\partial p_\alpha}\frac{\partial p_\alpha}{\partial p^*} \tag{2.25}$$

respectively. The right-hand sides of these equations coincide if

$$\lambda^\alpha \frac{\partial \varphi_\alpha}{\partial p^*} = \frac{\partial h}{\partial p_\alpha}\frac{\partial p^\alpha}{\partial p^*}. \tag{2.26}$$

Using the equation (2.23), this condition may be rewritten as

$$\lambda^\alpha \left(\frac{\partial \varphi_\alpha}{\partial p^*} + \frac{\partial \varphi_\alpha}{\partial p_\beta}\frac{\partial p_\beta}{\partial p^*} \right) = 0. \tag{2.27}$$

This equality holds owing to the conditions of constraint $\varphi_\alpha = 0$. The variables p^* are treated analogously. Thus, the statement is proved.

A change of choice of the subsidiary conditions is equivalent to a canonical transformation in the space $\Gamma^{*2(n-m)}$ and therefore does not influence the physics of the problem. For quantization of the system Γ the independent variables p^*, q^* may be used. Then the evolution operator is given by the path integral

$$\int \exp\left\{ i \int [p^* \zeta^* - h(p^*, q^*)] dt \right\} \prod_t \frac{dp^* dq^*}{(2\pi)}, \tag{2.28}$$

where the initial and final values of the coordinates q^* are fixed. It is therefore desirable to be able to work directly in terms of the generalized system Γ. It is easy to verify that the path integral

$$\int \exp\left\{i\int [p_i\dot{q}_i - h(p,q) - \lambda^\alpha\varphi^\alpha(p,q)]dt\right\} \times \prod_{t,\alpha}\delta(\chi^\alpha)\prod_t \det|\{\varphi_\alpha,\chi_\beta\}|\prod_t \frac{dp\,dq}{2\pi}\frac{d\lambda}{2\pi} \tag{2.29}$$

coincides with the integral (2.28). Indeed, integrating over λ, one can rewrite the formula (2.29) in the form

$$\int \exp\left\{i\int [p_i\dot{q}_i - h(p,q)]dt\right\} \times \prod_{t,\alpha}\delta(\chi^\alpha)\delta(\varphi^\alpha)\prod_t \det|\{\varphi_\alpha,\chi_\beta\}|\prod_t \frac{dp\,dq}{2\pi}. \tag{2.30}$$

In terms of the variables $p^\alpha, q^\alpha, p^*, q^*$ the factor

$$\prod_t \delta(\varphi_\alpha)\delta(\chi_\alpha)\det|\{\varphi_\alpha,\chi_\beta\}| \tag{2.31}$$

is rewritten as

$$\prod_t \delta(\varphi_\alpha)\delta(q_\alpha)\det\left|\frac{\partial\varphi_\alpha}{\partial p_\beta}\right| = \prod_t \delta(q_\alpha)\delta[p_\alpha - p_\alpha(p^*,q^*)]. \tag{2.32}$$

As a result, the integral (2.29) is reduced by integration over p_α and q_α to (2.28).

Comparison of the formulas (2.5)-(2.8) and (2.9), (2.10) shows that the Yang-Mills field is indeed a generalized Hamilton system. We shall now apply the procedure just described to its quantization.

It is clear that in this case the gauge condition must play the role of the subsidiary condition. As such a condition we shall choose the relation

$$\partial_k A_k = 0. \tag{2.33}$$

This condition is admissible. Indeed, it is obvious that

$$\{\partial_k A_k^a(x), \partial_i A_i^b(y)\} = 0. \tag{2.34}$$

Further

$$\{C^a(x), \partial_k A_k^b(y)\} = -\partial_k[\partial_k\delta^{ab} - gt^{abc}A_k^c(x)]\delta(x-y). \tag{2.35}$$

The operator $M_C = \Delta\delta^{ab} - gt^{abc}A_k^c(c)_k$ is reversible within the framework of perturbation theory. The inverse operator M_C^{-1} is an integral operator, the kernel $M_C^{-1}(x,y)$ of which is defined by the integral equation

$$M_C^{-1ab}(x,y) = \frac{1}{4\pi}\frac{\delta^{ab}}{|x-y|} + g\int\frac{dz}{4\pi}t^{acd}\frac{A_k^c(z)}{|x-z|}\partial_k M_C^{-1ab}(z,y) \tag{2.36}$$

and may be calculated by iterations as a formal series in g. (Notice that for large fields A_k the operator M_C may have a nonzero eigenvalue, so that M_C^{-1} will cease to exist. This problem, however, is beyond the scope of perturbation theory, and we shall not discuss it here.)

The form of the subsidiary condition (2.33) suggests that in order to find the coordinates q^*, it is convenient to use the orthogonal expansion $\mathcal{A}_k$,

$$\mathcal{A}_k = \mathcal{A}_k^L + \mathcal{A}_k^T \tag{2.37}$$

in longitudinal and transverse components. Here,

$$\mathcal{A}_K^L = \partial_k \mathcal{B}(x); \quad \mathcal{B}(x) = \frac{1}{4\pi} \int \frac{1}{|x-y|} \partial_k \mathcal{A}_k(y) dy, \tag{2.38}$$

where

$$\partial_k \mathcal{A}_k^T = 0. \tag{2.39}$$

It is clear that the transverse components $\mathcal{A}_k^T(x)$ play the role of q^*. The momenta that are their conjugates are the transverse components $\mathcal{E}_k^T(x)$. The equation for the constraints is the equation for the longitudinal part $\mathcal{E}_k^L(x)$. If one puts

$$\mathcal{E}_k^T(x) = \partial_k \mathcal{Q}(x), \tag{2.40}$$

then the equation for the constraint will be written as

$$\Delta \mathcal{Q} - g[\mathcal{A}_k, \partial_k \mathcal{Q}] - g[\mathcal{A}_k, \mathcal{E}_k^T] = 0, \tag{2.41}$$

where the operator M_C, already known to us, takes part. This equation allows us to express the longitudinal component $\mathcal{E}_k^L$ in terms of $\mathcal{E}_k^T$ and $\mathcal{A}_k^T$. After substituting the solution here into the Hamiltonian $h(\mathcal{A}, \mathcal{E})$, we obtain the Hamiltonian $h^*(\mathcal{A}^T, \mathcal{E}^T)$ in the form of an infinite series in the constant g. The variables $\mathcal{A}^T$, $\mathcal{E}^T$ and the Hamiltonian h^* are the true Hamiltonian variables for the Yang-Mills field. The field configuration $\mathcal{A}^T$ at fixed time t is given by two functions of x. This means that the Yang-Mills field has two possible states of polarization.

We can now write down the S-matrix for the Yang-Mills field in terms of the path integral:

$$\begin{aligned} S = \lim_{\substack{t'' \to \infty \\ t' \to -\infty}} \int \exp \Bigg\{ \int d^3k \frac{1}{2} \left[\sum_{i=1,2} a_i^{*b}(k,t'') a_i^b(k,t'') + a_i^{*b}(k,t') a_i^b(k,t') \right] \\ + i \int_{t'}^{t''} dt \int d^3x \Bigg[\left(-\frac{1}{4} \right) \times \mathrm{tr}[\mathcal{E}_i^T(x,t) \dot{\mathcal{A}}_i^T(x,t)] - \dot{\mathcal{E}}_i^T(x,t) \mathcal{A}_i^T(x,t)] \\ - h^*(\mathcal{E}^T, \mathcal{A}^T) \Bigg] \Bigg\} \prod \frac{da_i^*(k,t) da_i(k,t)}{2\pi i}, \end{aligned} \tag{2.42}$$

where

$$A_l^{T,b}(x,t) = \frac{1}{(2\pi)^{3/2}} \sum_{i=1,2} \int \left[e^{ikx} a_i^b(k,t) u_l^i(k) + e^{-ikx} a_i^{*b}(k,t) u_l^i(k) \right] \frac{d^3k}{\sqrt{2\omega}},$$

$$E_l^{T,b}(x,t) = \frac{i}{(2\pi)^{3/2}} \sum_{i=1,2} \int \left[-e^{ikx} a_i^b(k,t) u_l^i(k) + e^{-ikx} a_i^{*b}(k,t) u_l^i(k) \right] \frac{\sqrt{\omega} d^3k}{\sqrt{2}} \tag{2.43}$$

and $u_l^i(k), i = 1,2$ are two polarization vectors, which may be represented by two arbitrary orthonormalized vectors orthogonal to the vector k. Here we assume that the asymptotic conditions

$$a^{*b}(k,t'') \underset{t''\to\infty}{\longrightarrow} e^{i\omega t''} a_i^{*b}(k); \qquad a_i^b(k,t') \underset{t'\to-\infty}{\longrightarrow} e^{-i\omega t'} a_i^b(k) \tag{2.44}$$

are fulfilled.

This formula is not very convenient for the construction of the diagram technique, since the Hamiltonian h^* is known only in the form of a series in the constant g; and besides, it generates vertices that are nonlocal with respect to the space coordinates. Of course, this is only a technical difficulty, but it strongly impedes practical calculations, in particular the construction of a renormalization procedure. This deficiency vanishes if we use the representation for the S-matrix in the form of an integral over all functions $\mathcal{A}_i(k,t), \mathcal{E}_i(k,t)$:

$$\begin{aligned} S = \lim_{\substack{t''\to\infty \\ t'\to-\infty}} \int \exp\Bigg\{ \int d^3k \Bigg[\frac{1}{2} \sum_{i=1,2} a_i^{*b}(k,t'') a_i^b(k,t'') \\ + a_i^{*b}(k,t') a_i^b(k,t') \Bigg] + i \int_{t'}^{t''} dt \int d^3x \times \left(-\frac{1}{4} \right) \operatorname{tr}[\mathcal{E}_l(x,t) \dot{\mathcal{A}}_l(x,t) \\ - \dot{\mathcal{E}}_l(x,t) \mathcal{A}_l(x,t) - \mathcal{E}_l^2(x,t) - \mathcal{G}_l^2(x,t) + 2\mathcal{A}_0(\partial_l \mathcal{E}_l - g[\mathcal{A}_l, \mathcal{E}_l])] \Bigg\} \\ \times \prod_{x,t} \delta(\partial_l \mathcal{A}_l) \det M_C[\mathcal{A}] \prod_{x,t} d\mathcal{A}_l d\mathcal{E}_l d\mathcal{A}_0. \end{aligned} \tag{2.45}$$

Here the boundary terms $a_i^*(k,t''), a_i(k,t)$ are defined by the same formulas as before, that is, in terms of the transverse fields $\mathcal{A}_i^T$. We can now integrate over the momenta $\mathcal{E}_k$, taking into account the boundary conditions, as we did in Chapter 2 in the case of a scalar field. As a result, we obtain for the normal symbol of the S-matrix the expression

$$\begin{aligned} S = N^{-1} \int \exp\left\{ i \int dx \left[\frac{1}{8} \operatorname{tr} \mathcal{F}_{\mu\nu} \mathcal{F}_{\mu\nu} \right] \right\} \\ \times \prod_x \delta(\partial_k \mathcal{A}_k) \prod_t \det M_C[\mathcal{A}] \prod_x d\mathcal{A}_\mu \end{aligned} \tag{2.46}$$

where integration is performed over all fields $\mathcal{A}_\mu(x)$, the asymptotic behavior of their three-dimensional parts,

$$\mathcal{A}_l^T(x)_{t\to\pm\infty} \to \mathcal{A}_{l,\,\substack{\text{out}\\ \text{in}}}^T(x); \tag{2.47}$$

$$A_{l,\,\substack{\text{in}\\ \text{out}}}^{T,b} = \frac{1}{(2\pi)^{3/2}} \sum_{i=1,2} \int \left[a_{i,\,\substack{\text{in}\\ \text{out}}}^{b}(k) e^{ikx-i\omega t} u_l^i(k) + a_{i,\,\substack{\text{in}\\ \text{out}}}^{*b}(k)^{-ikx+i\omega t} u_l^i(k) \right] \frac{d^3k}{\sqrt{2k_0}},$$

$$a_{i,\text{in}}^b(k) = a_i^b(k), \quad a_{i,\text{out}}^{*b}(k) = a_i^{*b}(k), \quad i = 1, 2,$$

being fixed, and the definition of the quadratic form of the action being correspondingly supplemented.

In the formula (2.46), as in the case of the scalar field, the Feynman functional $\exp\{i\times \text{action}\}$ is integrated. However, integration is not performed over all fields. The integration measure contains explicitly the δ-function of the gauge condition. This is a manifestation of the relativity principle, according to which it is necessary to integrate not over all fields, but only over classes of gauge-equivalent fields; the δ-function selects one representative from each class, and the determinant provides the correct normalization of the integration measure. The asymptotic conditions are also in accordance with the choice of the gauge condition.

Expansion of the integral (2.46) in a perturbation-theory series gives rise to the diagram technique. The propagator is defined by the Gaussian integral

$$I(J) = N^{-1} \int \exp\left\{ i \int dx \,\mathrm{tr} \left[\frac{1}{8} (\partial_\nu \mathcal{A}_\mu - \partial_\mu \mathcal{A}_\nu)^2 - \frac{1}{2} \mathcal{Y}_\mu \mathcal{A}_\mu \right] \right\} \prod_x \delta(\partial_k \mathcal{A}_k) d\mathcal{A}_\mu \tag{2.48}$$

with the Feynman boundary conditions for $\mathcal{A}_k^T$. This integral is equal to

$$\exp\left\{ \frac{i}{2} \int J_\mu^a(x) D_{\mu\nu}^c(x-y) J_\nu^a(y) dx\, dy \right\} \tag{2.49}$$

where D_{uv}^c is the propagator sought:

$$\begin{aligned} D_{ml}^c(x) &= \frac{1}{(2\pi)^4} \int e^{-ikx} \left(\delta^{ml} - \frac{k_m k_l}{|k|^2} \right) (k^2 + i0)^{-1} dk, \\ D_{m0}^c(x) &= D_{0m}^c(x) = 0; \\ D_{00}^c(x) &= -\frac{1}{(2\pi)^4} \int e^{-ikx} \frac{1}{|k|^2} dk. \end{aligned} \tag{2.50}$$

For a proof we shall use the integral representation of the δ-function (2.5.18). Then $I(J)$ is given by the Gaussian integral,

$$\begin{aligned} I(J) = N^{-1} \int \exp\Big\{ i \int dx \Big[-\frac{1}{4} (\partial_\nu A_\mu^l - \partial_\mu A_\nu^l)^2 \\ + J_\mu^l A_\mu^L + \lambda^l \partial_k A_k^l \Big] \Big\} \prod_x dA_\mu d\lambda, \end{aligned} \tag{2.51}$$

for the calculation of which it is necessary to find the extremum of the exponent. The equations

$$\begin{aligned} &\partial_\nu(\partial_\nu A_k^l - \partial_k A_\nu^l) + J_k^l + \partial_k \lambda^l = 0, \\ &\partial_\nu(\partial_\nu A_0^l - \partial_0 A_\nu^l) + J_0^l = 0; \qquad \partial_k A_k^l = 0 \end{aligned} \tag{2.52}$$

are rewritten as

$$\begin{aligned} &\Box A_k^l + (\partial \lambda^l - \partial_0 \partial_k A_0^l) + J_k^l = 0, \\ &\Delta A_0^l - J_0^l = 0; \qquad \partial_k A_k^l = 0 \end{aligned} \tag{2.53}$$

and have unique solutions under the above-formulated boundary conditions. It may be assumed also that the source J satisfies the transversality condition

$$\partial_k J_k^l = 0. \tag{2.54}$$

As a result, the solution is given by the formula

$$A_\mu^l(x) = \int D_{\mu\nu}^c(x-y) J_\nu^l(y) dy, \tag{2.55}$$

where $D_{\mu\nu}^c(x)$ is the Coulomb propagator just introduced.

The explicit expression for $D_{\mu\nu}^c$ shows that only three-dimensionally transverse components $\mathcal{A}_\mu$ propagate in time, in agreement with our boundary conditions.

A defect of the diagram technique in the Coulomb gauge is the lack of explicit relativistic invariance. In the next section we shall show that in the integral (2.46) defining the S-matrix, one can pass to the manifestly covariant gauge.

To conclude this section we shall give a description of the alternative Hamiltonian formulation of the Yang-Mills theory, using the gauge condition $\mathcal{A}_0 = 0$. This gauge is an improvement on the Coulomb one in that it is also admissible beyond the scope of perturbation theory. Let us show that in each class of gauge equivalent fields there exists a field satisfying the condition

$$\mathcal{A}_0 = 0. \tag{2.56}$$

For this we point ut that the equation

$$\frac{\partial}{\partial}\omega(x,t) = -\omega(x,t)\mathcal{A}_0(x,t) \tag{2.57}$$

allows a solution of the form

$$\omega_0(x,t) = T\exp\left\{-\int_{-\infty}^{t} \mathcal{A}_0(x,s)ds\right\}, \tag{2.58}$$

where the symbol T signifies that the exponential is to be ordered in time. From the equation (2.57) it follows that

$$\mathcal{A}_\mu^{\omega_0} = \omega_0 \mathcal{A}_\mu \omega_0^{-1} + \partial_\mu \omega_0 \omega_0^{-1} \tag{2.59}$$

satisfies the condition

$$\mathcal{A}_0^{\omega_0} = 0. \tag{2.60}$$

In addition to the matrix $\omega_0(x)$, an analogous property is possessed by the matrices $\omega(x)$ of the form

$$\omega(x) = \omega(x)\omega_0(x), \tag{2.61}$$

where $\omega(x)$ is an arbitrary matrix from Ω, depending only on space coordinates. Thus, the Hamilton gauge does not completely abolish gauge arbitrariness in the definition of the Yang-Mills field, but reduces the gauge group to the group of matrices $\omega(x)$.

We shall now show that the equations of motion in the gauge $\mathcal{A}_0 = 0$ are actually Hamiltonian. For this it is convenient to use the equations of motion, formulated as first-order equations, following from the Lagrangian (2.1):

$$\begin{aligned} \partial_\nu \mathcal{A}_\mu - \partial_\mu \mathcal{A}_\nu + g[\mathcal{A}_\mu, \mathcal{A}_\nu] - \mathcal{F}_{\mu\nu} = 0, \\ \partial_\mu \mathcal{F}_{\mu\nu} - g[\mathcal{A}_\mu, \mathcal{F}_{\mu\nu}] = 0. \end{aligned} \tag{2.62}$$

Consider these equations in the three-dimensional formulation. In the notation $\mu = (0, k), \nu = (0, l)$, etc., the ten equations (2.61) are rewritten as

$$\begin{aligned} \partial_0 \mathcal{F}_{0k} &= \partial_l \mathcal{F}_{lk} - g\,[\mathcal{A}_l, \mathcal{F}_{lk}]\,, \\ \partial_0 \mathcal{A}_k &= \mathcal{F}_{k0} \\ \mathcal{F}_{ik} &= \partial_k \mathcal{A}_i - \partial_i \mathcal{A}_k + g\,[\mathcal{A}_l, \mathcal{A}_k] \\ \mathcal{C}(x) &= \partial_k \mathcal{F}_{0k} - g\,[\mathcal{A}_k, \mathcal{F}_{0k}] = 0 \end{aligned} \tag{2.63}$$

Eliminating the variables $\mathcal{F}_{ik}$ with the help of the equations of motion, we see that the system of equations (2.63) has an explicitly Hamiltonian form,

$$\begin{aligned} \partial_0 E_k^a(x,t) &= -\frac{\delta H}{\delta A_k^a(x,t)} = \{H, E_k^a(x,t)\} \\ \partial_0 A_k^a(x,t) &= \frac{\delta H}{\delta E_k^a(x,t)} = \{H, A_k^a(x,t)\}\,, \\ H &= \int h d^3x, \end{aligned} \tag{2.64}$$

where the above-introduced notation for E_k, h, and the Poisson bracket is used. The last equation of (2.63),

$$\mathcal{C}(x,t) = 0 \tag{2.65}$$

represents an equation of constraint. As we have already seen, the Poisson bracket $\{H, \mathcal{C}(x,t)\}$ vanishes,

$$\{H, \mathcal{C}(x,t)\} = \partial_0 \mathcal{C}(x,t) = 0 \tag{2.66}$$

so that $\mathcal{C}(x,t)$ generates an infinite set of integrals of motion.

We shall show that $\mathcal{C}(x)$ are generators of infinitesimal gauge transformations, remaining after the imposition of the gauge condition $\mathcal{A}_0 = 0$. To do this, we associate with an arbitrary matrix $\alpha(x)$ in the adjoint representation $\Omega(x)$ the quantity $C(\alpha)$:

$$C(\alpha) = -\frac{1}{2}\mathrm{tr}\left\{\int C(x)\alpha(x)d^3x\right\}. \tag{2.67}$$

The commutation relations (2.7) in this notation are rewritten as

$$\{C(\alpha), C(\beta)\} = gC([\alpha, \beta)] \tag{2.68}$$

This shows that $C(\alpha)$ defines the Lie-algebra representation of the group of gauge transformations, consisting of matrices $\alpha(x)$. The action of this representation on the variables $A_1(x)$ and $\mathcal{E}_1(x)$ is given by the formula

$$\begin{aligned} \delta A_l &= \{C(\alpha), A_l(x)\} = \partial_l\alpha(x) - g\,[A_l(x), \alpha(x)]\,, \\ \delta\mathcal{E}_k &= \{C(\alpha), \mathcal{E}_k(x)\} = -g\,[\mathcal{E}_k(x), \alpha(x)]\,. \end{aligned} \tag{2.69}$$

Thus, indeed, the $C(\alpha)$ are generators of gauge transformations, remaining in the Hamilton gauge.

In accordance with the relativity principle the observables $0(A_k, \mathcal{E}_k)$ are gauge-invariant and therefore must commute with $C(\alpha)$. This condition is a system of first-order differential equations, for which the relation (2.68) plays the role of the integrability condition and expresses one of the six functions $A_k, \mathcal{E}_k$, upon which 0 depends, in terms of all the others. Together with the conditions of constraint (2.65), this reduces the number of independent functions to four, in agreement with the calculated number of degrees of freedom in the Coulomb gauge.

Let us see how this classical picture is transferred to the quantum case. In the operator formulation the Hamiltonian, the constraint C, and the observables 0 become operators, that satisfy the relations

$$\begin{aligned} &\frac{1}{i}[C(\alpha), C(\beta)] = gC([\alpha, \beta]), \\ &[H, C(\alpha)] = 0; \qquad [O, C(\alpha)] = 0. \end{aligned} \tag{2.70}$$

We cannot directly equate the operator C to zero, although the formulas (2.70) show that there exists a subspace, formed by the vectors ψ satisfying the equation

$$C\psi = 0, \tag{2.71}$$

and that this subspace is invariant with respect to the operators corresponding to the observables. The condition (2.71) replaces the classical equation $C = 0$, and the constructed subspace is the true space of states of our physical system.

Thus, our task consists in constructing the scattering matrix describing transition from the asymptotic state ψ' to the state ψ''. Unlike the case of Coulomb gauge, when the asymptotic states include only three-dimensional transverse quanta, now the space of asymptotic states also contains longitudinal quanta, together with the transverse ones. It may be realized as the tensor product of the spaces of transverse and logitudinal quanta: $\mathcal{H} = \mathcal{H}^t \bigotimes \mathcal{H}^L$. as $\mathcal{H}^T$ one can choose the Fock space for transverse quanta. At the same time, from equations (2.63) it can be seen that the longitudinal field component in the absence of interaction satisfies the equation of free motion, instead of the equation for a harmonic oscillator. Therefore for $\mathcal{H}^L$ it is natural to choose the coordinate or momentum representation. Below we shall adopt the notation $\psi^L \equiv |\mathcal{E}_L>$, if $\mathcal{H}^L$ is realized in the momentum representation, and $\psi^L \equiv |A_L>$, if $\mathcal{H}^L$ is realized in the coordinate representation.

Physical states are determined by the condition (2.71). In the case of asymptotic states this condition is simplified. Usually it is assumed that the Gauss law is linearized asymptotically:

$$\lim_{|t|\to\infty} e^{iH_0t}C(\alpha)e^{-iH_0t} = C_0(\alpha) + O(1), \tag{2.72}$$

where

$$C_0(\alpha) = -\frac{1}{2}\int \operatorname{tr} \partial_k\mathcal{E}_k(x,t)\alpha(x)d^3x \tag{2.73}$$

is a generator of linearized gauge transformations:

$$\delta\varepsilon_k = 0, \qquad \delta A_k = \partial_k\alpha(x), \tag{2.74}$$

and H_0 is the free operator for the energy:

$$H_0 = -\frac{1}{4}\operatorname{tr}\int [\mathcal{E}_k^2 + (\partial_k A_l - \partial_l A_k)^2]\, d^3x. \tag{2.75}$$

The assumption concerning the linearization is standard for perturbation theory and is equivalent to the hypothesis of the interaction being switched off adiabatically. For our purposes a weaker condition is sufficient, namely, the linearization of the Gauss law in the three-dimensional transverse field components. Here, however, we shall not touch upon this point, and the condition (2.72) will be considered satisfied.

It can be readily seen that physical states annihilated by the operator C_0 can be represented in the form

$$|\Psi\rangle = |\Psi^T\rangle \otimes |0\rangle \equiv |\Psi^T, 0\rangle, \tag{2.76}$$

where $|\Psi^T\rangle$ denotes an arbitrary Fock vector in the space of transferse states, and $|0\rangle$ is the eigenvector of the operator $\mathcal{E}^L$ with an eigenvalue equal to zero.

The operator C_0 commutes with the S-matrix. This is demonstrated by the following formal computation:

$$\begin{aligned}
SC_0 &= \lim_{\substack{t''\to+\infty\\ t'\to-\infty}} e^{iH_0t''}e^{-iH(t''-t')}e^{-iH_0'}C_0 \\
&= \lim_{\substack{t''\to+\infty\\ t'\to-\infty}} e^{iH_0t''}e^{-iH(t''-t')}Ce^{-iH_0t'} \\
&= \lim_{\substack{t''\to+\infty\\ t'\to-\infty}} e^{iH_0t''}Ce^{-iH(t''-t')}e^{-iH_0t'} \\
&= \lim_{\substack{t''\to+\infty\\ t'\to-\infty}} C_0e^{iH_0t''}e^{-iH(t''-t')}e^{-iH_0t'} = C_0S.
\end{aligned} \tag{2.77}$$

Hence it folows that the S-matrix is unitary in the physical subspace, i.e.,

$$\langle \Psi^T, \mathcal{E}^L | S | \bar{\Psi}^T, 0 \rangle \sim \delta(\mathcal{E}^L). \tag{2.78}$$

Since vectors of the physical subspace belong to the continuous spectrum, the transition matrix element between two physical states, $\langle \Psi^T, 0 | S | \bar{\Psi}^T, 0 \rangle$, is singular. However, taking advantage of the property (2.78), it can be redefined in a natural manner:

$$\langle \Psi^T, 0 | S | \bar{\Psi}^T, 0 \rangle \stackrel{\text{def}}{=} \int \prod_x d\mathcal{E}^L(x) e^{\frac{i}{2} \operatorname{tr} \gamma \int \mathcal{E}_k^L(x) \mathcal{E}_k^L(x) dx} \langle \Psi^T, \mathcal{E}^L | S | \bar{\Psi}^T, 0 \rangle, \tag{2.79}$$

where γ is an arbitrary constant. Since the matrix element in the right-hand side is proportional to $\delta(\mathcal{E}^L)$, this expression is independent of γ.

To calculate the transition amplitudes of interest, we note that an arbitrary matrix element may, in accordance with formula (3.29) of Chapter 2, be written in the form

$$\begin{aligned} &\langle \Psi^T, \mathcal{E}^L | S | \bar{\Psi}^T, \mathcal{E}^L \rangle \\ &\quad = \exp\left\{ -i \int H_{\text{int}} \left(\frac{1}{i} \frac{\delta}{\delta J} \right) dx \right\} \langle \Psi^T, \mathcal{E}^L | S_0(J) | \bar{\Psi}^T, \bar{\mathcal{E}}^L \rangle, \end{aligned} \tag{2.80}$$

where

$$\begin{aligned} &\langle \Psi^T, \mathcal{E}^L | S_0(J) | \bar{\Psi}^T, \bar{\mathcal{E}}^L \rangle \\ &\quad = \lim_{\substack{t'' \to \infty \\ t' \to -\infty}} \langle \Psi^T, \mathcal{E}^L | e^{iH_0 t''} U_0(t'', t') e^{-iH_0 t'} | \bar{\Psi}^T \bar{\mathcal{E}}^L \rangle, \end{aligned} \tag{2.81}$$

and $U_0(t'', t')$ is the evolution operator for scattering on an external source $\int J_i^a(x) A_i^a(x) dx$.

The matrix element (2.81) determines the propagator of the Feynman diagram technique, knowledge of which makes possible calculation of the S-matrix of interest by the formula (2.80). In the formula (2.81) the dependence on the longitudinal and on the transverse components factorizes and the problem is reduced to the independent computation of $\langle \Psi^T | S(J^T) | \bar{\Psi}^T \rangle$ and $\langle \mathcal{E}^L | S(J^L) | \bar{\mathcal{E}}^L \rangle$. The transverse matrix element was calculated above. Thus, for instance,

$$\langle 0 | S_0(J^T) | 0 \rangle = \exp\left\{ \frac{i}{2} \int J_i^{Ta}(x) D_{ij}^{abC}(x-y) J_j^{bT}(y) dx\, dy \right\}, \tag{2.82}$$

where D_{ij}^c is the Coulomb propagator (2.50). The longitudinal matrix element $\langle \mathcal{E}^L | S_0(J^L) | 0 \rangle$ can be represented in the form

$$\begin{aligned} \langle \mathcal{E}^L | S_0(J^L) | 0 \rangle = &\lim_{\substack{t'' \to \infty \\ t' \to -\infty}} \int e^{-i \int dx E^{La} A^{La}} \langle A^L | e^{iH_0 t''} | A_2^L \rangle \\ &\times \langle A_2^L | U_0^L(t'', t') | A_1^L \rangle \langle A_1^L | e^{-iH_0 t'} | 0 \rangle \prod dA^L dA_1^L dA_2^L. \end{aligned} \tag{2.83}$$

In this formula we have taken advantage of the property of completeness of the set of vectors $|A^L\rangle$.

Each of the matrix elements occurring here is expressed through a functional integral of the form

$$\int_{\substack{\mathcal{A}^L(x,t_2)=\mathcal{A}_i^L(x) \\ \mathcal{A}^L(x,t_1)=\mathcal{A}_j^L(x)}} \prod_x d\mathcal{A}^L(x) \exp\left\{ i \int_{t_1}^{t_2} \mathcal{L}(x,t)dx\,dt \right\}, \tag{2.84}$$

where

$$\mathcal{L}(x,t) = \frac{1}{2}(\partial_0 A_i^{a,L})^2 + J_i^{a,L}(x)A_i^{a,L}(x). \tag{2.85}$$

(For the first and third matrix elements $J_i^L = 0$.)

The integral (2.84) is Gaussian, so it equals the integrand calculated at the extremum point of the exponent. The extremum values of $\mathcal{A}^L$ are determined from the solution of the corresponding classical equations and are given by the formulas:

$$A_k^{a,L}(x,t) = \frac{1}{2}\int_{t_1}^{t_2} |t-s| J_k^{a,L}(x,s)ds + C_k^a(x)t + D_k^a(x)$$

$$C_k^a(x) = \frac{1}{t_2 - t_1}(A_{k,i}^{a,L} - A_{k,j}^{a,L}) + \int_{t_1}^{t_2} s j_k^{a,L}(x,s)ds$$

$$-\frac{1}{2}(t_1 + t_2)\int_{t_1}^{t_2} J_k^{a,L}(x,s)ds, \tag{2.86}$$

$$D_k^a(x) = a_{k,j}^{a,L} - \frac{1}{2}\int_{t_1}^{t_2}(s - t_1)J_k^{a,L}(x,s)ds - C_k^a(x)t_1.$$

Substituting these expressions into (2.83) and performing trivial integration over $\mathcal{A}^L$ we obtain the expression for the matrix element $\langle \mathcal{E}^L | S_0(J^L)|0\rangle$. To obtain the transition matrix element between physical states, it is necessary, in accordance with the definition (2.79), to integrate this expression with the weight $\exp\left\{\frac{i\gamma}{2}\int E_k^{La}(x)E_k^{La}(x)dx\right\}$. As a result, we obtain

$$\langle 0|S_0(J^L)|0\rangle = \exp\left\{\frac{1}{2}\int J_k^{a,L}(x)D_{ka}^{ab,L}(x,y)J_n^{b,L}(y)dx\,dy\right\}, \tag{2.87}$$

where

$$D_{kn}^{ab,L}(x,y) = \bar{D}_{kn}^{ab,L}(\mathbf{x}-\mathbf{y})D(x_0,y_0), \tag{2.88}$$

$$\bar{D}_{kn}^{ab,L}(\mathbf{x}-\mathbf{y}) = \frac{\delta^{ab}}{(2\pi)^3}\int dk e^{ik(\mathbf{x}-\mathbf{y})}\frac{k_k k_n}{k^2}, \tag{2.89}$$

$$D(x_0,y_0) = \frac{1}{2}|x_0 - y_0| - \frac{1}{2}|x_0 + y_0| + \gamma. \tag{2.90}$$

If in the formula (2.79) integration were performed over the initial, instead of the final, states, we would obtain for the function $D(x_0, y_0)$

$$D(x_0,y_0) = \frac{1}{2}|x_0 - y_0| + \frac{1}{2}|x_0 + y_0| + \gamma. \tag{2.91}$$

Combining the results for the transverse and logitudinal matrix elements we obtain the final expression for the Green function of the Yang-Mills field in the gauge $\mathcal{A}_0 = 0$:

$$D_{in}^{ab}(x,y) = -\frac{\delta^{ab}}{(2\pi)^4} \int \frac{dk e^{-ik(x-y)}}{k^2 + i0} \left(\delta_{in} - \frac{k_i k_n}{k^2}\right)$$
$$+ \left[\frac{1}{2}|x_0 - y_0| \pm \frac{1}{2}(x_0 + y_0) + \gamma\right] \times \frac{\delta^{ab}}{(2\pi)^3} \int dk e^{ik(x-y)} \frac{k_i k_n}{k^2}. \quad (2.92)$$

As can be seen, the propagator (2.92) does not depend only on the difference $(x - y)$. At first sight this violates invariants with respect to translations. Gauge-invariant quantities, however, are readily shown to exhibit translational invariance. At any rate, this follows from the fact that terms that are not invariant with respect to translations can be discarded by gauge transformation.

Unlike the diagram technique inthe Coulomb gauge, in the gauge $\mathcal{A}_0 = 0$ there exist no vertices generated by det M_C, which simplifies somewhat the form of the interaction. However, the absence of manifest translational invariance strongly hinders practical calculations and represents a serious obstacle to carrying out consistently the program of renormalizations, to be discussed in thenext chapter.

The final expression for the S-matrix can be written in the form of a path integral of the relativistic-invariant action by formally introducing integration over $\mathcal{A}_0$:

$$S = N^{-1} \int \exp\left\{i \int dx \left[\frac{1}{8} \mathrm{tr} \mathcal{F}_{\mu\nu} \mathcal{F}_{\mu\nu}\right]\right\} \prod_x \delta(\mathcal{A}_0) d\mathcal{A}_\mu, \quad (2.93)$$

where the boundary conditions described above are assumed to hold for $\mathcal{A}_\mu$.

This integral also permits interpretation in the spirit of the relativity principle. It is in integral over the classes of gauge-invariant fields under another gauge condition, defining the choice of representatives.

3.3 Covariant Quantization Rules and the Feynman Diagram Technique

As was already pointed out, the expression for the S-matrix obtained in the previous section is not manifestly covariant. This is inconvenient for performing calculations within the framework of perturbation theory, especially for renormalization procedures. The path-integral method allows us to get rid of this defect. The relativity principle suggests that for this it is necessary to pass to a relativistic-invariant parametrization of the classes of gauge-equivalent fields, that is, to choose a relativistic-invariant gauge. The most simple relativistic-invariant gauge condition is the Lorentz condition,

$$\partial_\mu \mathcal{A}_\mu = 0. \quad (3.1)$$

We shall show how to pass to the Lorentz gauge starting with the already known expression for the S-matrix in the Coulomb gauge (2.46). From a geometric point of view we must transfer the measure, defined on the surface $\Phi_C \equiv \partial_k A_k = 0$, to the surface $\Phi_L \equiv \partial_\mu A_\mu = 0$ along the trajectories of the gauge group. Formally this may be achieved in the following way. We introduce the functional $\Delta_L(\mathcal{A})$, proceeding from the condition

$$\Delta_L(\mathcal{A}) \int \prod_x \delta(\partial_\mu \mathcal{A}_\mu^\omega) d\omega = 1 \tag{3.2}$$

where integration is performed over the measure $\prod_x d\omega(x)$ and $d\omega$ is the invariant measure on the group Ω:

$$d(\omega\omega^0) = d(\omega^0\omega) = d\omega. \tag{3.3}$$

The functional $\Delta_L(\mathcal{A})$ is obviously gauge-invariant:

$$\Delta_L(\mathcal{A}^\omega) = \Delta_L(\mathcal{A}), \tag{3.4}$$

which follows directly from the invariance of the integration measure.

Using the relation (3.2), it is possible to rewrite the expression for the S-matrix (2.46) as

$$S = N^{-1} \int \exp\left\{ i \int dx \left[\frac{1}{8} \operatorname{tr} \mathcal{F}_{\mu\nu}\mathcal{F}_{\mu\nu} \right] \right\} \prod_x \delta(\partial_k \mathcal{A}_k)$$
$$\times \prod_t \det M_C(\mathcal{A}) \Delta_L(\mathcal{A}) \prod_x \delta(\partial_\mu \mathcal{A}_\mu^\omega) d\omega d\mathcal{A}. \tag{3.5}$$

Now note that the functional $\prod_t \det M_C(\mathcal{A})$ coincides with the gauge-invariant functional $\Delta_C(\mathcal{A})$ on the surface $\Phi_C \equiv \partial_k \mathcal{A}_k = 0$, where $\Delta_C(\mathcal{A})$ is introduced analogously to $\Delta_L(\mathcal{A})$:

$$\Delta_C(\mathcal{A}) \int \delta(\partial_k \mathcal{A}_k^\omega) d\omega = 1. \tag{3.6}$$

Indeed, if $\mathcal{A}_k$ satisfies the condition $\partial_k \mathcal{A}_K = 0$, then $\omega = 1$ is obviously the root of the argument of the δ-function (the only one within the framework of perturbation theory). Therefore in the integral (3.6) it is sufficient to integrate only in the vicinity of the unit element. For $\omega(x) \approx 1 + u(x)$ we have

$$\partial_k \mathcal{A}_k^\omega = \Delta u - g\left[\mathcal{A}_k(x), \partial_k u(x)\right] = M_C u(x) \tag{3.7}$$

and

$$\prod_x d\omega(x) = \prod_x du(x). \tag{3.8}$$

Thus the integral is calculated explicitly, and we obtain that

$$\Delta_C(\mathcal{A})|_{\partial_k \mathcal{A}_k = 0} = \prod_x \det M_C(\mathcal{A}). \tag{3.9}$$

Let us go back to our integral (3.5), in which, as just shown, we can put

$$\prod_x \delta(\partial_k \mathcal{A}_k) \prod_x \det M_C = \prod_x \delta(\partial_k \mathcal{A}_k) \Delta_C . \tag{3.10}$$

We perform a change of variables,

$$\mathcal{A}_\mu \to \mathcal{A}_\mu^{\omega^{-1}}, \tag{3.11}$$

the Jacobian of which is obviously equal to unity. Owing to the invariance of the action and factors Δ_L, Δ_C, the integral (3.5) can be rewritten in the form

$$S = N^{-1} \int \exp\left\{ i \int dx \left[\frac{1}{8} \operatorname{tr} \mathcal{F}_{\mu\nu} \mathcal{F}_{\mu\nu} \right] \right\} \prod_x \delta(\partial_\mu \mathcal{A}_\mu) \Delta_L(\mathcal{A})$$

$$\times \delta(\partial_k \mathcal{A}_k^{\omega^{-1}}) \Delta_C(\mathcal{A}) d\omega d\mathcal{A}. \tag{3.12}$$

Substituting ω in the integral over $d\omega$ for ω^{-1} and using the formula (3.6), we see that the last two factors in the integrand of (3.12) may be dropped. As a result, we obtain the expression for the S-matrix in the Lorentz gauge:

$$S = N^{-1} \int \exp\left\{ i \int dx \left[\frac{1}{8} \operatorname{tr} \mathcal{F}_{\mu\nu} \mathcal{F}_{\mu\nu} \right] \prod_x \Delta_L(\mathcal{A}) \delta(\partial_\mu \mathcal{A}_\mu) d\mathcal{A}. \right. \tag{3.13}$$

Reasoning entirely analogous to that which led to (3.9) shows that on the surface $\Phi_L \equiv \partial_\mu \mathcal{A}_\mu(x) = 0$ the functional Δ_L is equal to

$$\Delta_L(\mathcal{A})|_{\partial_\mu \mathcal{A}_\mu = 0} = \det M_L , \tag{3.14}$$

where the operator M_L is defined by the formula

$$M_L \alpha(x) = \Box \alpha - g \partial_\mu [\mathcal{A}_\mu, \alpha] = \Box \alpha + W(\mathcal{A}) \alpha \tag{3.15}$$

We recall that we have already encountered the determinants $\det M_C$ and $\det M_L$ which appear here, in the first chapter, while we were formulating the admissibility of the gauge condition.

We have not yet discussed the influence of a change of variables on the asymptotic conditions in the integral (2.46). Therefore the formula (3.13) for the S-matrix is as yet somewhat formal. In particular, our reasoning does not make it clear what meaning is to be attributed to the determinant of M_L. The point is that for a consistent definition of the operator M_L in the whole space of variables x, we need boundary conditions as $t \to \pm\infty$. In another manner this problem may be formulated as follows. For defining the determinant it is natural to use the formula

$$\det M_L = \exp\left\{ \operatorname{Tr} \ln M_L \right\} = \exp\left\{ \operatorname{Tr} \ln \Box + \operatorname{Tr} \ln(1 + \Box^{-1} W(\mathcal{A})) \right\} . \tag{3.16}$$

Here the symbol Tr stands for the operation of taking the trace, including also integration over the coordinates.

The first factor is an insignificant constant that changes only the normalization constant N. The second factor generates an additional term in the action, which has the form

$$\begin{aligned}\operatorname{Tr}\ln(1+\Box^{-1}W(\mathcal{A})) &= \sum_n \frac{(-1)^{n+1}}{n}\operatorname{Tr}(\Box^{-1}W)^n \\ &= -\frac{g^2}{2}\int dx_1 dx_2 \operatorname{tr}\{\mathcal{A}_\mu(x_1)\mathcal{A}_\mu(x_2)\}\,\partial_\mu D(x_1-x_2) \\ &\quad\times \partial_\mu D(x_2-x_1) - \ldots + (-1)^{n+1}\frac{g^n}{n}\int dx_1\ldots dx_n \\ &\quad\times \operatorname{tr}\{\mathcal{A}_{\mu_1}(x_1)\ldots\mathcal{A}_{\mu_n}(x_n)\} \\ &\quad\times \partial_{\mu_1}D(x_1-x_2)\ldots\partial_{\mu_n}D(x_n-x_1) - \ldots, \end{aligned} \tag{3.17}$$

where $D(x)$ is the Green function of the d'Alembertian operator. This Green function is not defined uniquely, and the question arises which boundary conditions must be imposed for its unique definition. In practice, it is a question of how to get around the pole in the integral

$$D = -\frac{1}{(2\pi)^4}\int \frac{e^{-ikx}}{k^2}dk \tag{3.18}$$

defining the Green function. An analogous problem arises in defining the Green function $D^L_{\mu\nu}(x-y)$ corresponding to the quadratic form in the Lorentz gauge. The formal answer, obtained by inverting this quadratic form, is

$$D^L_{\mu\nu}(x-y) = \frac{1}{(2\pi)^4}\int e^{-ik(x-y)}\left\{g^{\mu\nu} - \frac{k_\mu k_\nu}{k^2}\right\}\frac{1}{k^2}dk. \tag{3.19}$$

In this formula it is also necessary to clarify in which sense one gets around the poles of the integrand.

For an answer to the question on the boundary conditions, it is necessary to perform a transformation of the integral (2.46) into the integral (3.13) before passing to the time limit $t'' \to \infty, t' \to -\infty$. We recall that in the Coulomb gauge, besides the boundary conditions on the three-dimensional components of the potential $\mathcal{A}^T_k$, there exists the condition

$$\partial_k \mathcal{A}_k = 0, \tag{3.20}$$

which is satisfied in the whole interval $t' \le t \le t''$, including $t = t'', t = t'$. The change of the variables

$$\mathcal{A}_\mu \to \mathcal{A}^{\omega^{-1}}_\mu = \omega^{-1}\mathcal{A}_\mu\omega + \partial_\mu\omega^{-1}\omega \tag{3.21}$$

should not violate this condition.

The gauge functions $\omega(x,t)$ is conveniently chosen to satisfy the following boundary conditions:

$$\omega(x,t'') = 1, \qquad \omega(x,t') = 1. \tag{3.22}$$

Such choice provides for the boundary values of the transverse components $\mathcal{A}^T_k$ remaining unchanged as $t' \to -\infty, t'' \to +\infty$, since in this limit the transformation (3.21) is reduced to the substitution

$$\mathcal{A}_k \to \mathcal{A}_k, \qquad \mathcal{A}_0 \to \mathcal{A}_0 - \partial_0 \alpha, \tag{3.23}$$

where

$$\omega = \exp\{\alpha\} \tag{3.24}$$

Thus, the formal definition of the operator M_L, given by the formula (3.15), must be supplemented by the boundary conditions

$$\alpha(x, t'') = \alpha(x, t') = 0. \tag{3.25}$$

The Green function D, appearing in the expansion of the determinant in the perturbation-theory series, is the Green function of the d'Alembertian operator with the same boundary conditions. Such a function has the form

$$D_1(x, y) = \frac{1}{(2\pi)^3} \int e^{ik(x-y)} \frac{\sin\left[|k|(x_0 - t')\right] \sin\left[|k|(y_0 - t'')\right]}{|k| \sin\left[|k|(t'' - t)\right]} d^3k, \tag{3.26}$$

$$x_0 \leq y_0;$$

at $x_0 \geq y_0 D_1(x, y)$ is determined by the symmetry condition.

$$D_1(x, y) = D_1(y, x).$$

With such a definition of the operator M_L, its determinant is positive in the framework of perturbation theory, and this justifies its use in the formula (3.14) instead of det M_L.

The problem of getting around the poles in the Green function $D^L_{\mu\nu}$ is solved in an analogous manner. For its definition at finite t', t'' we must solve the equations

$$\Box \mathcal{A}_\mu = \mathcal{Y}_\mu, \qquad \partial_\mu \mathcal{A}_\mu = 0, \tag{3.27}$$

where $\mathcal{Y}_\mu$ satisfies thecompatibility condition

$$\partial_\mu \mathcal{Y}_\mu = 0. \tag{3.28}$$

The boundary conditions for the system are

$$a_i^*(k, t'') = a_i^*(k) e^{i\omega t''}, \qquad a_i(k, t') = a_i(k) e^{-i\omega t'} \tag{3.29}$$

$$(i = 1, 2),$$

$$\partial_k \mathcal{A}_k(x, t) = 0 \qquad t = t', \quad t = t''$$

The boundary conditions for $\mathcal{A}_0$ follow from the system (3.27) itself and have the form

$$\partial_0 \mathcal{A}_0 = 0, \qquad t = t' \quad t = t''. \tag{3.30}$$

The solution of the system (3.27) has the form

$$\mathcal{A}_i^T(x) = \mathcal{A}_i^T(x) + \int \tilde{D}(x, y) \mathcal{Y}_i^T(y) dy, \tag{3.31}$$

where

$$A_l^{bT_0}(x) = \sum_{i=1,2} \frac{1}{(2\pi)^{3/2}} \int e^{ikx - i\omega t} a_i^b(k) u_l^i(k) + e^{-ikx + i\omega t} a_i^{b*}(k) u_l^i(k) \frac{d^3k}{\sqrt{2\omega}}, \tag{3.32}$$

and the vectors $u_l^i(k)$ are the ones introduced previously in (2.43). The Green function $\tilde{D}(x,y)$ has the form

$$\tilde{D}(x,y) = D_c(x-y)\theta(t''-y_0)\theta(y_0-t'), \tag{3.33}$$

and as $t' \to -\infty, t'' \to +\infty$, transforms into the causal Green function $D_c(x-y)$. The remaining components $\mathcal{A}_l^L(x)$ and $\mathcal{A}_0(x)$ are given by the formulas

$$\mathcal{A}_0(x) = \int D_2(x,y)\mathcal{Y}_0(y)dy; \qquad \mathcal{A}_l^L = \int D_2(x,y)\mathcal{Y}_l^L(y)dy, \tag{3.34}$$

where $D_2(x,y)$ is the Green function of the d'Alembertian operator with the boundary conditions

$$\partial_0 \alpha|_{t=t''} = \partial_0 \alpha|_{t=t'} = 0. \tag{3.35}$$

This function has the form

$$D_2(x,y) = \frac{1}{(2\pi)^3} \int e^{ik(x-y)} \frac{\cos\left[|k|(x_0-t')\right] \cos\left[|k|(y_0-t'')\right]}{|k| \sin\left[|k|(t''-t')\right]} d^3k, \tag{3.36}$$

$$x^0 \le y^0.$$

At $x_0 \ge y_0$, $D_2(x,y)$ is defined by the symmetry condition. Combining the formulas (3.33), (3.36), and (3.26), we obtain the Green function in the Lorentz gauge for a finite time interval consistent with the Coulomb boundary conditions.

Let us now try to pass to the limit as $t'' \to \infty, t' \to -\infty$ in the expressions obtained. A limit for the function $D(x,y)$ exists, and it coincides with the causal Green function $D_c(x-y)$. This is in agreement with the fact that the three-dimensionally transverse components of $\mathcal{A}_l^T(x)$ correspond to physical polarizations.

The functions $D_1(x,y)$ and $D_2(x,y)$ have no limits as $t'' \to +\infty, t' \to -\infty$. At the same time the limit of the integral (3.13), defining the S-matrix in the Lorentz gauge, must exist, since by construction this integral is equal to the Coulomb integral (2.46), for which a limit exists. This means that under expansion of the S-matrix (3.13) in the perturbation-theory series the total contribution of the function D_1 and D_2 tends to a definite limit. Formally, the simplest way to calculate this limit is to regularize the functions D_1 and D_2 in the same manner, for example, by adding an infinitesimal imaginary part to the integration variable k^2. As a result of such a regularization, the oscillating exponentials in the integrands for D_1 and D_2 will become either increasing or decreasing at large $|t'|, |t''|$, and the limit will exist. It is most convenient to assume that k^2 has a negative imaginary part $-i0$, since in this case the limits of the functions D_1 and D_2 coincide with the causal function $D_c(x)$, and for the full Green function in the Lorentz gauge we obtain the manifestly covariant expression,

$$D^L_{\mu\nu}(x) = -\frac{1}{(2\pi)^4}\int \left(g^{\mu\nu} - \frac{k_\mu k_\nu}{k^2+i0}\right)\frac{1}{k^2+i0}e^{-ikx}dk. \tag{3.37}$$

At the same time, the Green function figuring in the expansion of the determinant det M_L in the perturbation-theory series also become causal, and the determinant itself becomes a complex-valued functional of $\mathcal{A}_\mu$.

We may once more emphasize here that the specific regularization used herein is not the only possible one. For instance, substituting $k^2 \to k^2 + i0$, we would obtain antichronological Green functions for nonphysical polarizations, and the imaginary part of the determinant would change sign. The Green function $D^L_{\mu\nu}$, however, would then lose its manifest covariance.

The rather lengthy reasoning given has led us to the following answer to the question put above: All circuits of the poles of Green functions may be considered to be Feynman ones; that is, we must interpret $1/k^2$ as $(k^2+i0)^{-1}$. Thus, the S-matrix in the Lorentz gauge has the form

$$S = N^{-1}\int_{\substack{\mathcal{A}_\mu \to \mathcal{A}_{\mu\,\text{in, out}} \\ t\to\mp\infty}} \exp\left\{i\int \frac{1}{8}\text{tr}\, \mathcal{F}_{\mu\nu}\mathcal{F}_{\mu\nu}dx\right\} \times \prod_x \Delta_L(\mathcal{A})\delta(\partial_\mu \mathcal{A}_\mu)d\mathcal{A}, \tag{3.38}$$

where $\mathcal{A}_{\mu\,\text{in, out}}$ is the solution of the equations

$$\Box \mathcal{A}_\mu = 0, \qquad \partial_\mu \mathcal{A}_\mu = 0, \tag{3.39}$$

parametrized by the amplitudes $a_\mu(k)$ and $a^*_\mu(k)$ such that

$$a_0 = 0, \qquad k_l a_l = 0; \qquad a^*_0 = 0, \qquad k_l a^*_l = 0, \tag{3.40}$$

the amplitude $a_1(k)$ being given in $\mathcal{A}_{\text{in}}$ (positive frequency wave) and $a^*_1(k)$ in $\mathcal{A}_{\text{out}}$ (negative frequency wave).

An analogous derivation of the formula (3.13) for the S-matrix could be given, proceeding from the Hamilton gauge $\mathcal{A}_0 = 0$.

The obtained formula (3.38) is not the only possible relativistic-invariant expression for the S-matrix. Integration over gauge-equivalent classes may be performed in ways other than by selecting a representative from each class with the help of the gauge condition. Analysis of our passage from the Coulomb gauge to the Lorentz gauge reveals that in the formula (3.2) it is not necessary to use a functional of the δ-function type as an integrand. Instead, one may take any functional $B(\mathcal{A})$ which is not gauge-invariant, but for which the integral

$$\Delta_B^{-1}(\mathcal{A}) = \int [B(\mathcal{A}^\omega_\mu)]\prod_x d\omega. \tag{3.41}$$

converges. As a result, a path integral for the S-matrix appears in which $\delta(\partial_\mu\mathcal{A}_\mu)det M_L(\mathcal{A})$ is replaced by $\Delta_B(\mathcal{A}_\mu)B(\mathcal{A}_\mu)$. Taking the functional

$$\exp\left\{-\frac{i}{4\alpha}\int \text{tr}\,(\partial_\mu\mathcal{A}_\mu)^2dx\right\}, \tag{3.42}$$

for $B(\mathcal{A})$, we obtain a family of free Green functions $D^{\alpha}_{\mu\nu}$,

$$D^{\alpha}_{\mu\nu}(x) = -\frac{1}{(2\pi)^4}\int e^{-ikx}\left\{g_{\mu\nu} - \frac{k_\mu k_\nu(1-\alpha)}{k^2+i0}\right\}\frac{1}{k^2+i0}, \tag{3.43}$$

which contains the most widely used special cases: at $\alpha = 0$ we come back to the Lorentz gauge, and at $\alpha = -1$ we obtain the diagonal Green function.

We shall give some formal reasoning that realizes this program as simply as possible. First, let us pass from the Lorentz gauge to the generalized Lorentz gauge:

$$\partial_\mu \mathcal{A}_\mu(x) = a(x), \tag{3.44}$$

where $a(x)$ is an arbitrary matrix, using the same reasoning as when passing from the Coulomb to the Lorentz gauge. The corresponding functional $\Delta_a(\mathcal{A})$, is given by the formula

$$[\Delta_a(\mathcal{A})]^{-1} = \int \prod_x \delta[\partial_\mu \mathcal{A}^\omega_\mu - a(x)]d\omega, \tag{3.45}$$

coincides on the surface

$$\partial_\mu \mathcal{A}_\mu = a(x) \tag{3.46}$$

with the functional $\det M$, where the operator M is given by the formula (3.15). Thus the generating functional (3.38) for the S-matrix is identically rewritten as

$$S = N^{-1}\int_{\mathcal{A}\to\mathcal{A}_{\substack{\text{in}\\\text{out}}}} \exp\left\{i\int\frac{\text{tr}}{8}\mathcal{F}_{\mu\nu}\mathcal{F}_{\mu\nu}dx\right\} \times \prod_x \delta(\partial_\mu \mathcal{A}_\mu - a(x)) \det M d\mathcal{A}. \tag{3.47}$$

Since the initial functional does not depend on a, we can integrate it over $a(x)$ with the weight,

$$\exp\left\{-i\frac{\text{tr}}{4\alpha}\int a^2(x)dx\right\}, \tag{3.48}$$

which leads only to a change in the normalization constant N. Performing the integration, we obtain the generating functional for the S-matrix in the form

$$S = N^{-1}\int_{\mathcal{A}\to\mathcal{A}_{\substack{\text{in}\\\text{out}}}} \exp\left\{i\int \text{tr}\left[\frac{1}{8}\mathcal{F}_{\mu\nu}\mathcal{F}_{\mu\nu} - \frac{1}{4\alpha}(\partial_\mu \mathcal{A}_\mu)^2\right]dx\right\} \times \prod_x \det M d\mathcal{A}. \tag{3.49}$$

Extending the notion of the gauge condition, we shall call this functional the S-matrix in the α-gauge.

Expansion of this functional in the perturbation-theory series generates the diagram technique with the Green functions (3.43). In order to make this reasoning quite rigorous it is necessary, as above, to deal more carefully with the boundary conditions. We shall not do this here and shall restrict ourselves to pointing out that all the Green functions may be chosen to be causal. The equivalence of the

S-matrix in various gauges will be discussed in greater detail in the next chapter, in connection with the problem of renormalization.

It is possible to introduce gauges of an even more generalized form, for which the longitudinal part of the Green function of the Yang-Mills field is an arbitrary function of k^2. For this it is sufficient to use as the functional $B(A)$ an expression of the type $\exp\{-(i/4\alpha)\mathrm{tr}[f(\Box)\partial_\mu A_\mu]^2 dx\}$, where $f(\Box)$ is an arbitrary function of the d'Alembertian operator. All the reasoning given above for the case of $f \equiv 1$ is applicable without any change to this case. Gauges of this type will be used later in the discussion of regularization and renormalization.

The expression (3.50) for the S-matrix contains the nonlocal functional det M and therefore does not look like the familiar integral of the Feynman functional $\exp\{i \times \text{action}\}$ over all fields. We may, however, use for det M the integral representation

$$\det M = \int \exp\left\{i \int \bar{c}^a(x) M^{ab} c^b(x) dx\right\} \prod_x d\bar{c}\, dc, \tag{3.50}$$

where $\bar{c}(x)$ and $c(x)$ are anticommuting scalar functins (generators of the Grassman algebra), satisfying the emission condition

$$\begin{aligned} d^{a*}_{\text{out}}(k) &= 0, & g^{a*}_{\text{out}}(k) &= 0, \\ d^{a}_{\text{in}}(k) &= 0, & g^{a}_{\text{in}}(k) &= 0, \end{aligned} \tag{3.51}$$

where d, g, d^*, g^* are given by usual formulas

$$\begin{aligned} c^a(x,t)_{\substack{\text{in}\\ \text{out}}} &= \frac{1}{(2\pi)^{3/2}} \int [e^{ikx - i\omega t} d^a_{\substack{\text{in}\\ \text{out}}}(k) + e^{-ikx + i\omega t} g^{a*}_{\substack{\text{in}\\ \text{out}}}(k)] \frac{d^3k}{\sqrt{2\omega}}, \\ c^a(x,t)_{\substack{\text{in}\\ \text{out}}} &= \frac{1}{(2\pi)^{3/2}} \int [e^{ikx - i\omega t} g^a_{\substack{\text{in}\\ \text{out}}}(k) + e^{-ikx + i\omega t} d^{a*}_{\substack{\text{in}\\ \text{out}}}(k)] \frac{d^3k}{\sqrt{2\omega}}, \end{aligned} \tag{3.52}$$

Using this representation, we rewrite the formula (3.50) for S as

$$\begin{aligned} S = N^{-1} \int \exp\Big\{ i \int \mathrm{tr} \Big[&\frac{1}{8} \mathcal{F}_{\mu\nu} \mathcal{F}_{\mu\nu} - \frac{1}{4\alpha} (\partial_\mu A_\mu)^2 \\ &- \frac{1}{2} \bar{c}(\Box c - g\partial_\mu [A_\mu, c]) \Big] dx \Big\} \prod_x dA d\bar{c}\, dc, \end{aligned} \tag{3.53}$$

where as $t \to \pm\infty$

$$A_\mu \to A_{\mu\, \substack{\text{in}\\ \text{out}}}, \qquad c \to c_{\substack{\text{in}\\ \text{out}}}, \qquad \bar{c} \to \bar{c}_{\substack{\text{in}\\ \text{out}}}, \qquad c = c^q a t^a. \tag{3.54}$$

At the cost of introducing the fictitious fields $\bar{c}, c$ we have succeeded intaking the relativity principle into acount in such a way that the S-matrix is represented in the form of an integral of $\exp\{i \times \text{action}\}$, where the action is local and has a nondegenerate quadratic form, and integration is performed over all fields. This allows us to develop a perturburation theory for the functional (3.54), as was done inthe previous chapter for the case of a scalar field, proceeding from the Gaussian integral.

With this aim we introduce the generating functional for the Green functions

$$Z(J_\mu, \bar{\xi}, \xi) = N^{-1} \int \exp\left\{ i \int \mathcal{L}_\alpha(x) + [J^a_\mu A^a_\mu + \bar{\xi}^a c^a + \bar{c}^a \xi^a] dx \right\} \prod_x d\mathcal{A} d\bar{c}\, dc$$

$$= \exp\left\{ iV \left\{ \frac{1}{i}\frac{\delta}{\delta J_\mu}, \frac{1}{i}\frac{\delta}{\delta \xi}, \frac{1}{i}\frac{\delta}{\delta \bar{\xi}} \right\} \right\} \times \exp\left\{ \frac{i}{2} \int [J^a_\mu(x) D^{\alpha ab}_{\mu\nu}(x-y) J^b_\nu(y)\right.$$

$$\left. + 2\bar{\xi}^a(x) D^{ab}(x-y)\xi^b(y)]\, dx\, dy \right\}, \qquad (3.55)$$

where $J^a_\mu, \bar{\xi}^a, \xi^a$ are the sources of the fields $A^a_\mu, c^a, \bar{c}^a$, the ξ^a and $\bar{\xi}^b$ anticommuting with each other and with the fields c^{-a}, c^b, and

$$V(\mathcal{A}_\mu, \bar{c}, c) = \frac{1}{8} \mathrm{tr} \int \{2g(\partial_\nu \mathcal{A}_\mu - \partial_\mu \mathcal{A}_\nu)[\mathcal{A}_\mu, \mathcal{A}_\nu]$$

$$+ g^2([\mathcal{A}_\mu, \mathcal{A}_\nu])^2 + 4g\bar{c}\partial_\mu[\mathcal{A}_\mu, c]\} dx; \qquad (3.56)$$

the derivatives with respect to $\bar{\xi}$ are considered left-handed, and with respect to ξ right-handed. In the integral (3.56) all the integration variables satisfy the Feynman boundary conditions. The expansion of the functional Z in the perturbation-theory series generates the diagram technique. We list its elements, using from the beginning the momentum representation.

1. The propagator of the vector particles:

$$= D^{ab}_{\mu\nu}(p) = -\frac{\delta^{ab}}{p^2 + i0}\left(g_{\mu\nu} - \frac{p_\mu p_\nu}{p^2 + i0}(1-\alpha) \right). \qquad (3.57)$$

2. The self-interaction of vector particles:

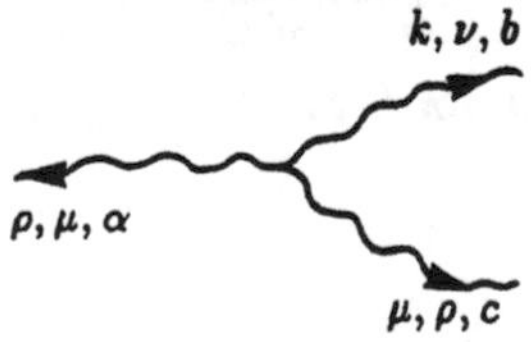

$$= V_{A^3}$$

$$= igt_{abc}[(p-k)_\rho g_{\mu\nu} + (k-q)_\mu g_{\nu\rho} + (q-p)_\nu g_{\mu\rho}]. \qquad (3.58)$$

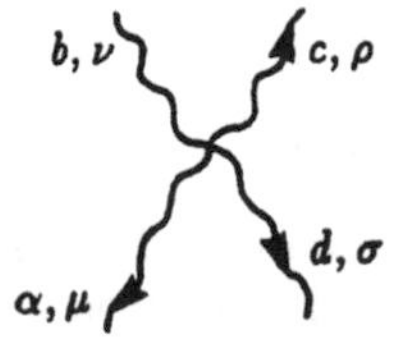

$$= V_{A^3} = g^2\{t_{abe}t_{cde}(g_{\mu\rho}g_{\nu\sigma} - g_{\mu\sigma}g_{\nu\rho})$$

$$+ t_{ace}t_{bde}(g_{\mu\nu}g_{\rho\sigma} - g_{\mu\sigma}g_{\rho\nu} + t_{ade}t_{cbe}(g_{\mu\rho}g_{\sigma\nu} - g_{\mu\nu}g_{\sigma\rho})\}. \qquad (3.59)$$

3. The propagator of the fictitious c-particles:

$$= D^{ab} = \frac{\delta^{ab}}{p^2 + i0} \qquad (3.60)$$

4. The interaction vertex of the fictitious c-particles with the Yang-Mills field:

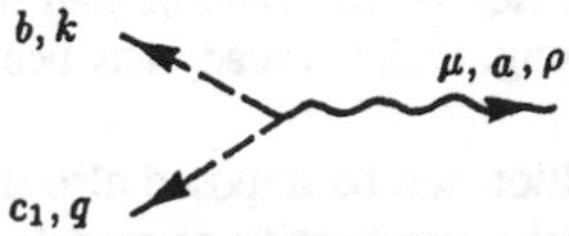

$$= V_{\bar{c}cA} = -igt_{abc}k_\mu. \tag{3.61}$$

Each diagram involving these elements defines the contribution to the Green functions $G_{n,m}(k_1, \dots, k_n | p_1, \dots, p_m)$ with n external legs for the vector particles and m legs for the fictitious c-particles. The contribution of a given diagram enters with the factor

$$\frac{1}{r}\left(\frac{i}{(2\pi)^4}\right)^{l-V}(-1)^s, \tag{3.62}$$

where V is the number of vertices, l is the number of internal lines, r is the order of the symmetry group of the diagram, and s is the number of closed loops of fictitious particles.

The S-matrix is calculated with the Green functions, using the reduction formulas:

$$S_{i_1 \dots i_n, j_1 \dots j_m}(k'_1, \dots, k'_n, k_1, \dots, k_m)$$
$$= k_1'^2 \dots k_n'^2 k_1^2 \dots k_m^2 \theta(k'_{10}) \dots \theta(k'_{n0}) \times \theta(-k_{10}) \dots \theta(-k_{m0}) u^{i_1}_{\mu_1} \dots u^{i_n}_{\mu_n}$$
$$\times G_{\mu_1 \dots \mu_n, \nu_1 \dots \nu_m}(k'_1 \dots k'_n, k_1 \dots k_m) u^{j_1}_{\nu_1} \dots u^{j_m}_{\nu_m}\Big|_{\substack{k_i^2 = 0 \\ k_j'^2 = 0}} \tag{3.63}$$

In this cumbersome formula we have multiplied each external vector line with a momentum k by k^2 and the polarization vector $u^i_\mu = (0, u^i_l), i = 1; 2$, and then passed to the mass shell $k^2 = 0$, assuming that $k_{i0} > 0$ for each incoming particle and $k_{i0} < 0$ for each outgoing particle. The fictitious particles have no corresponding external lines and enter into the S-matrix only by means of closed loops.

3.4 Interaction with Fields of Matter

The consideration of matter fields $\psi(x)$ interacting with the Yang-Mills field $\mathcal{A}_\mu(x)$ does not give rise to new difficulties in the quantization problem. The action of the gauge group on the field $\mathcal{A}_\mu(x)$ is described by the same formulas as without the matter fields. Therefore, the gauge condition, imposed only on the field $\mathcal{A}_\mu(x)$, fixes the choice of the representatives in the classes of gauge-equivalent fields $\mathcal{A}_\mu(x), \psi(x)$. This means that in the definition of the S-matrix for these fields, one may in the corresponding path integral integrate over the fields ψ with a measure already calculated beforehand (for instance, $\prod_x d\varphi(x)$ for a scalar field, $\prod_x d\bar{\psi}(x)d\psi(x)$ for a spinor field), and as a measure for the fields $\mathcal{A}_\mu$ take one of the measures calculated in the previous section for the Yang-Mills field in vacuum. A rigorous derivation must be based on the Hamiltonian formulation of the dynamics, and only repeats the reasoning, which already has been presented more than once.

At the same time the gauge condition can be imposed also on the matter field ψ. This is convenient, specifically, for the quantization of models with spontaneous symmetry breaking. An example of such a condition will be given below.

We shall start with the example of interacting Yang-Mills and spinor fields. The Lagrangian

$$\mathcal{L} = \frac{1}{8g^2}\operatorname{tr}\{\mathcal{F}_{\mu\nu}\mathcal{F}_{\mu\nu}\} + i\bar{\psi}\gamma_\mu\nabla_\mu\psi - m\bar{\psi}\psi \tag{4.1}$$

is invariant under the gauge transformation

$$\psi(x) \to \Gamma(\omega)\psi(x), \quad \mathcal{A}_\mu \to \mathcal{A}_\mu^\omega = \omega\mathcal{A}_\mu\omega^{-1} + \partial_\mu\omega\omega^{-1}. \tag{4.2}$$

The Hamilton gauge condition

$$\mathcal{A}_0 = 0 \tag{4.3}$$

is admissible and leads to the equations of motion in the generalized Hamiltonian formulation (with a natural modification taking account of the anticommutativity of the fields $\bar{\psi}, \psi$)

$$\begin{aligned} \partial_0\mathcal{A}_k &= \frac{\delta h}{\delta\mathcal{F}_{k0}} = \{h, \mathcal{A}\}, \quad \partial_0\mathcal{F}_{k0} = -\frac{\delta h}{\delta\mathcal{A}_k} = \{h, \mathcal{F}_{k0}\}, \\ \partial_0\psi &= -i\frac{\delta h}{\delta\psi^*} = \{h, \psi\}, \quad \partial_0\psi^* = i\frac{\delta h}{\delta\psi} = \{h, \psi^*\}, \end{aligned} \tag{4.4}$$

where

$$h = \int\left[i\bar{\psi}\gamma_k(\partial_k - \mathcal{A}_k)\psi + m\bar{\psi}\psi - \frac{1}{g^2}\operatorname{tr}(\mathcal{F}_{0k}^2 + \mathcal{F}_{ik}^2)\right]d^3x. \tag{4.5}$$

In addition, among the equations of motion there are the constraints

$$C^a(x) = \partial_k F_{k0}^a - t^{abc}A_k^b F_{0k}^c + i\bar{\psi}\gamma_0\Gamma(T^a)\psi, \tag{4.6}$$

differing from (2.63) by the last term, which is constructed in terms of the matter fields. Note that thisterm is the 0-component of the current

$$J_\mu^a = \bar{\psi}\gamma_\mu\Gamma(T^a)\psi, \tag{4.7}$$

which is conserved in the absence of interaction. The relations

$$\begin{aligned} \{C^a(x), C^b(y)\} &= t^{abc}\delta(x-y)C^c(x), \\ \{h, C^a\} &= 0, \\ \{C^a(x), A_k^b(y)\} &= \delta^{ab}\partial_k\delta(x-y) - t^{acb}A_k^c\delta(x-y), \\ \{C^a(x), \psi(y)\} &= \Gamma(T^a)\psi(x)\delta(x-y), \\ \{C^a(x), \bar{\psi}(y)\} &= -\Gamma(T^a)\bar{\psi}(x)\delta(x-y), \end{aligned} \tag{4.8}$$

analogous to (2.66), (2.68) and (2.69) show that $C^a(x)$ is a generator of gauge transformations, which remain after the imposition of the gauge condition $\mathcal{A}_0 = 0$. The parameters α^a of this transformation are independent of x_0.

The fields of matter enter quadratically into $C^a(x)$, so that the constraint is linearized on the solution of the free equations as $t \to \infty$,

$$C^a(x) \underset{|t|\to\infty}{\longrightarrow} C_0^a(x), \tag{4.9}$$

where

$$C_0^a = \partial_k F_{k0}^a. \tag{4.10}$$

As a result, repeating the reasoning of Section 3.2, we come to the conclusion that in the quantum case, if the S-matrix is constructed with the Hamiltonian H in the large space where all the fields $A_k^a, D_k^a, \bar{\varphi}^a, \varphi^a$ act, then it commutes with the operator $C_0^a(x)$:

$$[S, C_0^a(x)] = 0. \tag{4.11}$$

In other words, in the presence of fields of matter only quanta of these fields and three-dimensionally transverse quanta of the Yang-Mills fields are scattered.

Note that these conclusions are based, as before, on the linearization of the constraint $C^a(x)$ at large values of time. In the framework of perturbation theory, such a linearization seems quite convincing, and we assume it to take place. At the same time we cannot exclude that beyond the scope of perturbation theory linearization does not occur. The models of quark confinement are based exactly on this circumstance.

Coming back to our S-matrix, we write it in the form of a path integral,

$$S = \int \exp\left\{ i \int_{\psi \to \psi_{\text{in, out}}} \mathcal{L}(x)dx \right\} \prod_x \delta(\mathcal{A}_0) d\mathcal{A} d\bar{\psi}\, d\psi, \tag{4.12}$$

where the boundary conditions formulated in Section 3.2 are implied for $\mathcal{A}$, and we apply to this integral the transformations already described in Section 3.3:

1. Integration over $\mathcal{F}_{k0}$.
2. Transfer to the generalized Lorentz gauge
$$\Phi_L = \partial_\mu \mathcal{A}_\mu(x) + a(x) = 0$$
using the formula
$$\delta(\mathcal{A}_0) d\mathcal{A}_\mu \to \Delta_L(\mathcal{A}) \delta(\partial_\mu \mathcal{A}_\mu + a) d\mathcal{A}_\mu; \tag{4.13}$$
3. Integration over the auxiliary function $a(x)$ with the Gaussian weight $\exp\left\{-(i/4\alpha)\mathrm{tr}\int a^2(x)dx\right\}$.

We shall obtain expressions both for the S-matrix and for the generating functional of the Green functions in the α-gauge, which differ from the formulas (3.53),

(3.55) only by the presence of fields $\bar{\varphi}, \varphi$ in the Lagrangian and in the terms with sources.

Besides the already introduced elements $G_{\mu\nu}, G, V_{A^3}, V_{A^4}, V_{\bar{c}cA}$, the diagram technique contains also a spinor line

p
$$= S = -\frac{m + p_\mu \gamma_\mu}{m^2 - p^2 - i0} \tag{4.14}$$

and vertex,

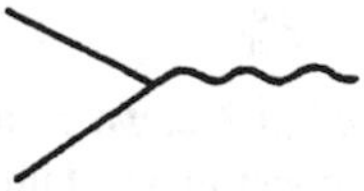
$$= V_{\bar{\psi}\psi A} = ig\gamma_\mu T(t_a). \tag{4.15}$$

Due to the already noted special features of integration over Fermi fields, each fermion cycle gives an additional factor (-1). In the reduction formulas the spinor legs are multiplied by $(2\pi)^{-3/2}\bar{u}_i(\gamma_\mu k_\mu - m)\theta(-k_0)$ for the incoming particle and by $(2\pi)^{-3/2}(\gamma_\mu k_\mu - m)u_i\theta(k_0)$ for the outgoing particle. For antiparticles u_i must be replaced by v_i and $\theta(k_0)$ by $\theta(-k_0)$.

The second example to be considered is the model with spontaneously broken symmetry. As the gauge group we choose the $SU(2)$ group and let φ be the scalar field inthe isospinor representation

$$\varphi = \begin{pmatrix} \varphi_1 \\ \varphi_2 \end{pmatrix}, \qquad \varphi^+ = (\varphi_1^*, \varphi_2^*). \tag{4.16}$$

As the starting point for quantization, we adopt the Lagrangian (3.32) of Chapter 1, which is obtained from (3.25, Chapter 1) upon imposition of the gauge condition (3.30):

$$\mathcal{L} = -\frac{1}{4}F^a_{\mu\nu}F^a_{\mu\nu} + \frac{m_1^2}{2}A^a_\mu A^a_\mu + \frac{1}{2}\partial_\mu\sigma\partial_\mu\sigma - \frac{1}{2}m_2^2\sigma^2$$
$$+\frac{m_1 g}{2}\sigma A^a_\mu A^a_\mu + \frac{g^2}{8}\sigma^2 A^a_\mu A^a_\mu - \frac{g m_2^2}{4m_1}\sigma^3 - \frac{g^2 m_2^2}{32 m_1^2}\sigma^4, \tag{4.17}$$
$$m_1 = \frac{\mu g}{\sqrt{2}}, \qquad m_2 = 2\lambda\mu.$$

Unlike the examples considered above, this Lagrangian is not degenerate, and for quantization we may apply the usual methods elaborated in Chapter 2 utilizing the example of the scalar field. Special attention must be paid only to the variable $A_0^\alpha(x)$ which in this case is neither a dynamic variable nor a Lagrangian factor. The equation of motion for A_0

$$\left(m_1 + \frac{g\sigma}{2}\right)^2 A_0 = \nabla_k \mathcal{F}_{0k} \tag{4.18}$$

permits expression of it through the independent dynamic variables A_k and σ and their conjugate momenta. Substituting the solution into the Lagrangian (4.17), we obtain a non-degenerate Lagrangian depending on these variables.

We already know that the quadratic form in the Hamiltonian determines the asymptotic conditions imposed on the variables of integration in the path integral for the S-matrix. In our case this quadratic form is of the form

$$h_0^* = \frac{1}{2}(F_{0k}^a)^2 + \frac{1}{2m_1^2}(\partial_k F_{0k}^a)^2 + \frac{1}{4}(\partial_i A_k^a - \partial_k A_i^a)^2 + \frac{m_1^2}{2}(A_k^a)^2 + \frac{1}{2}\sigma_0^2 + \frac{1}{2}(\partial_k \sigma)^2 + \frac{m_2^2}{2}\sigma^2 \tag{4.19}$$

and is diagonalized by the substitution

$$A_l^b(x) = (2\pi)^{-3/2}\sum_{i=1}^{3}\int (e^{ikx}a_i^b(k)e_l^i(k) + e^{-ikx}a_i^{*b}(k)e_l^i(k))\frac{d^3k}{\sqrt{2\omega_1}} \tag{4.20}$$

$$F_{0l}^b(x) = (2\pi)^{-3/2}\sum_{i=1}^{3}\int (e^{ikx}a_i^b(k)\tilde{e}_l^i(k) - e^{-ikx}a_i^{*b}(k)\tilde{e}_l^i(k))\frac{1}{i}\sqrt{\frac{\omega_1}{2}}d^3k, \tag{4.21}$$

where $e_l^1 = \tilde{e}_l^1$ and $e_l^2 = \tilde{e}_l^2$ are two arbitrary orthonormalized vectors orthogonal to the vector $\vec{k}$,

$$e_l^3 = \frac{k_l}{|k|}\frac{\omega_1}{m_1}; \quad \tilde{e}_l^3 = \frac{k_l}{|k|}\frac{m_1}{\omega_1}, \quad \omega_1 = \sqrt{k^2 + m_1^2}, \tag{4.22}$$

for the vector field, and by the standard substitution for the scalar field σ. The resultant free Hamiltonian assumes the form

$$\int h_0^* d^3x = \int d^3k \left(\sum_{i=1}^{3} a_i^{*b}a_i^b\omega_1 + a_\sigma^* a_\sigma \omega_2\right), \quad \omega_2 = \sqrt{k^2 + m_2^2}. \tag{4.23}$$

From the present computations the spectrum is seen to consist of three massive vector particles and one masive scalar particle.

The normal symbol of the S-matrix is given by the expression

$$S(A_l^{(0)}, \sigma^{(0)}) = \int_{\substack{A_l^a \to A^a_{i\binom{\text{in}}{\text{out}}} \\ \sigma \to \sigma_{\binom{\text{in}}{\text{out}}}}} \exp\left\{i\int \tilde{\mathcal{L}}(x)dx\right\}\prod_x dA_i(x)d\sigma(x). \tag{4.24}$$

where $A_i^{(0)}, \sigma^{(0)}$ are solutions of the free equations

$$A_l^{b(0)}(x) = \left(\frac{1}{2\pi}\right)^{3/2}\int [a_i^b(k)e_l^i(k)e^{-ikx} + a_i^{*b}(k)e_l^i(k)e^{ikx}]\frac{d^3k}{\sqrt{2\omega_1}}, \tag{4.25}$$

$$\sigma^{(0)}(x) = \frac{1}{(2\pi)^{3/2}}\int [a_\sigma(k)e^{-ikx} + a_\sigma^*(k)e^{ikx}]\frac{d^3k}{\sqrt{2\omega_2}}, \tag{4.26}$$

and the boundary conditions are given by the functions $A^a_{l\binom{\text{in}}{\text{out}}}, \sigma_{\binom{\text{in}}{\text{out}}}$ described by formulae similar to (4.25) and (4.26), involving the obvious substitution $a_i(k)$, $a_i^*(k) \to a_{i\binom{\text{in}}{\text{out}}}(k), a_{i\binom{\text{in}}{\text{out}}}(k)$, $a_\sigma(k), a_\sigma^*(k) \to a_{\sigma\binom{\text{in}}{\text{out}}}(k), a_{\sigma\binom{\text{in}}{\text{out}}}(k)$.

Here,

$$a^b_{i,\text{in}}(k) = a^b_i(k), \quad a^{*b}_{i,\text{out}}(k) = a^{*b}_i(k),$$
$$a_{\sigma,\text{in}}(k) = a_\sigma(k); \quad a^*_{\sigma,\text{out}}(k) = a^*_\sigma(k) = a^*_\sigma(k). \tag{4.27}$$

No conditions are imposed on the functions $a^b_{i,\text{out}}(k)$, $a^{*b}_{i,\text{in}}(k)$ and $a_{\sigma,\text{out}}(k)$, $a^*_{\sigma,\text{in}}(k)$.

As the Lagrangian $\tilde{\mathcal{L}}(x)$, one must take the result obtained by substitution of the solution of equations (4.18) into the formula (4.17).

A disadvantage of this formula is the absence of manifest relativistic invariance, due to integration being performed over only three components of the vector field. This defect can be readily removed by introducing integrtion over $\mathcal{A}_0$. Namely, the formula

$$S = \int_{\substack{\mathcal{A}\to\mathcal{A}_{\substack{\text{in}\\\text{out}}}\\ \sigma\to\sigma_{\substack{\text{in}\\\text{out}}}}} \exp\left\{i\int \mathcal{L}(x)dx\right\}\prod_x \left(m_1 + \frac{g}{2}\sigma\right)^2 d\mathcal{A}_\mu d\sigma, \tag{4.28}$$

where $\mathcal{L}(x)$ is determined by the formula (4.16), is equivalent to (4.24). Here,

$$\mathcal{A}^b_{0,\substack{\text{in}\\\text{out}}}(x) = \frac{1}{(2\pi)^{3/2}}\int \left[a^b_{i,\substack{\text{in}\\\text{out}}}(k)u^i_0(k)e^{-ikx} + a^{*b}_{i,\substack{\text{in}\\\text{out}}}(k)u^i_0(k)e^{ikx}\right]\frac{d^3k}{\sqrt{2\omega_1}}, \tag{4.29}$$

$$u^i_0 = 0, \quad i = 1,2, \quad u^3_0 = \frac{|k|}{m}, \tag{4.30}$$

and $\mathcal{A}^a_{l,\substack{\text{in}\\\text{out}}}$ and $\sigma_{\substack{\text{in}\\\text{out}}}$ have already been introduced above. Indeed, the integral (4.28) is reduced to (4.24) upon Gaussian integration over $\mathcal{A}_0$. The local factor $\left(m_1 + \frac{g}{2}\sigma(x)\right)^3$ is compensated by the corresponding determinant, while the asymptotic conditions (4.29) are matched with the equation (4.18).

We stress that $\partial_\mu \mathcal{A}_{\mu,\substack{\text{in}\\\text{out}}} = 0$ so the component $\partial_\mu \mathcal{A}_\mu$ of the vector field does not propagate, and we recall that the quadratic form in the action $\int \mathcal{L}(x)dx$ is defined as

$$\frac{1}{2}\int (\mathcal{A}_\mu - \mathcal{A}^{(0)}_\mu)(\Box + g_{\mu\nu}m_1^2)(\mathcal{A}_\mu - \mathcal{A}^{(0)}_\mu)dx, \tag{4.31}$$

where the operator $(\Box_{\mu\nu} + m_1^2 g_{\mu\nu})$ is supplied with Feynman boundary conditions.

The generating functional for the Green functions,

$$Z(J,\eta) = \int \exp\left\{i\int(\mathcal{L}(x) + J^a_\mu \mathcal{A}^a_\mu + \sigma\eta)dx\right\} \times \prod_x \left(m_1 + \frac{g}{2}\sigma\right)^3 d\mathcal{A}_\mu d\sigma \tag{4.32}$$

and the perturbation-theory diagram technique that follows from it contain some new features. First, the propagator of the vector field (4.51), which can be rewritten as

$$D^c_{\mu\nu}(x-y) = \left(\frac{1}{2\pi}\right)^4 \int e^{ik(x-y)}\left(g_{\mu\nu} - \frac{k_\mu k_\nu}{m_1^2}\right) \times \frac{1}{k^2 - m_1^2 + i0}d^4k, \tag{4.33}$$

has a higher degree of singularity at $x \sim y$, then the Green functions we have encountered until now. Indeed, its longitudinal part

$$\frac{k_\mu k_\nu}{m_1^2} \frac{1}{k^2 - m_1^2 + i0} \tag{4.34}$$

does not decrease at large k, so that its contribution to the propagator has a singularity of the power $\delta^{(4)}(x)$. Second, the integration measure contains the local factor $\det M_\varphi$. It may be formally written as

$$\det M_\Phi = \prod_x \left(m_1 + \frac{g}{2}\sigma(x)\right)^2 = \text{const} \cdot \exp\left\{V \int \ln\left(m_1 + \frac{g\sigma(x)}{2}\right)^3\right\} dx$$

$$= \text{const} \cdot \exp\left\{\delta^{(4)}(0)\left[-\sum_{n=1}^{\infty} \frac{(-1)^n}{n} \int \left(\frac{g\sigma(x)}{2m_1}\right)^n dx\right]\right\}, \tag{4.35}$$

$$V = \int dx = \delta^{(4)}(0).$$

In the framework of perturbation theory such an addition to the action generates new diagrams, the contribution of which is proportional to powers of $\delta^{(4)}(0)$ (of course, this expression is to be understood in the sense of a certain volume regularization). The role of these diagrams is to compensate the singular parts of other diagrams arising in the perturbation theory. Such singularities arise in the multiplication of δ-type contriutions ofvector particles to the Green functions.

Both indicated features of the diagram technique for the Lagrangian (4.58) show that it contains inconvenient singularities. Therefore it is more convenient to investigate the model under consideration in the Lorentz gauge or in the α-gauge, which can be introduced in a simple manner, using already familiar methods. The role of the above gauge (often called unitary) is that it gives the spectrum of particles and the asymptotic states of the model in a manifestly relativistic-invariant manner. In this sense it gives us a substitute for the Coulomb gauge of the Yang-Mills theory in vacuum.

Note that the normal symbol of the S-matrix can be written down utilizing the gauge invariant Lagrangian (1.3.25) if one returns to the complete set of fields $\mathcal{A}_\mu, \varphi$ and if the gauge condition is introduced into the path integral. To write the formula explicitly, we introduce the notation

$$\varphi_1 = \frac{iB_1 + B_2}{\sqrt{2}}; \qquad \varphi_2 = \mu\frac{1}{\sqrt{2}}(\sigma - iB_3), \tag{4.36}$$

and express the Lagrangian (1.3.25) through the fields $\mathcal{A}_\mu, \mathcal{B}, \sigma$. Then,

$$S = N^{-1} \int_{\substack{\mathcal{A} \to \mathcal{A}_{\text{in,out}} \\ \sigma \to \sigma_{\text{in,out}}}} \exp\left\{i \int \mathcal{L}(\mathcal{A}, \mathcal{B}, \sigma) dx\right\}$$

$$\times \prod_x \delta(\mathcal{B}) \left(m_1 + \frac{g}{2}\sigma(x)\right)^3 d\mathcal{A}_\mu d\mathcal{B} d\sigma, \tag{4.37}$$

where $\mathcal{L}(\mathcal{A}, \mathcal{B}, \sigma)$ results from substitution of (4.36) into (1.3.35).

It can readily be seen that the local factor $\prod_x (m_1 + \frac{g}{2}\sigma(x))^3$ is the Jacobian M_F of the gauge condition $B = 0$. Therefore, by repeating the procedure for transition to the covariant gauge, already described above, we obtain for the normal symbol of the S-matrix in the α-gauge the following expression:

$$S = N^{-1} \int_{\substack{A \to A_{\text{in,out}} \\ \sigma \to \sigma_{\text{in,out}}}} \exp\left\{ i \int \left(\mathcal{L}(x) + \frac{1}{2\alpha}(\partial_\mu A_\mu)^2 \right) dx \right\}$$

$$\times \det M_\alpha \prod_x dA_\mu dB d\sigma, \qquad (4.38)$$

where

$$\mathcal{L}(x) = -\frac{1}{4} F^a_{\mu\nu} F^a_{\mu\nu} + \frac{m_1^2}{2} A_\mu^2 + m_1 A^a_\mu \partial_\mu B^a + \frac{1}{2} \partial_\mu B^a \partial_\mu B^a$$

$$+ \frac{1}{2} \partial_\mu \sigma \partial_\mu \sigma - \frac{m_2^2}{2} \sigma^2 + \frac{g}{2} A^a_\mu (\sigma \partial_\mu B^a - B^a \partial_\mu \sigma - \epsilon^{abc} B^b \partial_\mu B^c)$$

$$+ \frac{m_1 g}{2} \sigma A_\mu^2 + \frac{g^2}{8} (\sigma^2 + B^2) A_\mu^2 - \frac{g m_2^2}{4 m_1} \sigma (\sigma^2 + B^2)$$

$$- \frac{g^2 m_2^2}{32 m_1^2} (\sigma^2 + B^2)^2 \qquad (4.39)$$

and

$$M_\alpha u = M u = \Box u - g \partial_\mu [A_\mu, u]. \qquad (4.40)$$

The asymptotic conditions for A^a_μ and σ are the same as before. The fields $\partial_\mu A^a_\mu$, B^a and the fictitious particles $\bar{c}^a, c^a$, taking part in the definition of $\det M$, do not propagate. In the construction of the diagram technique, it is convenient (but not necessary) to use the Feynman boundary conditions for them. The generatingfunctional for the Green functions is constructed in a standard way by means of the expression (4.38) for the S-matrix.

The diagram technique in the generalized α-gauge is somewhat cumbersome owing to the presence of mixed propagators. In calculations it is more convenient to use the Lorentz gauge $\alpha = 0$. In this gauge the diagram technique has the following elements:

1. A propagator corresponds to the line of a vector particle

$$-\delta^{ab} \left[\frac{g_{\mu\nu} - k_\mu k_\nu k^{-2}}{k^2 - m_1^2 + i0} \right]. \qquad (4.41)$$

2. The propagators of the fictitious particles $\bar{c}, c$ and their interaction vertices with a vector particle are the same as for the Yang-Mills theory in vacuum.

3. To lines of scalar particles B^a and σ there correspond the propagators

$$= \frac{\delta^{ab}}{k^2 + i0} \qquad (4.42)$$

$$= \frac{1}{k^2 - m_2^2 + i0}. \tag{4.43}$$

4. There exist numerous interaction vertices of the fields $\mathcal{A}_\mu, \mathcal{B}$ and σ, which are readily written out according to the Lagrangian (4.39).

The reduction formulas have the usual form, and we shall describe them in words. The external legs in the Green functions must be taken only for vector particles and for particles of the field σ. Each leg, corresponding to a vector field, is multiplied by $(k_j^2 - m_1^2)$ and by the polarization vector $u^i_\mu(k_j)$. Each leg, corresponding to the field σ, is multiplied by $(p_1^2 - m_2^2)$. Then it is necessary to pass on to the mass shell $k_j^2 = m_1^2, p_1^2 = m_2^2$, assuming k_0 and p_0 to be positive for outgoing particles and negative for incoming particles.

Another version of the diagram technique can be obtained if one chooses as the functional B in the formula (3.41), which removes degeneracy, the expression

$$\exp\left\{\frac{i}{2\alpha}\int(\partial_\mu A^a_\mu + \alpha m_1 B^a)^2 dx\right\}. \tag{4.44}$$

Computations entirely analogous to those which led us to the expression for the S-matrix in the α-gauge lead to the following result:

$$S = N^{-1}\int_{\substack{\mathcal{A}\to\mathcal{A}_{\text{in/out}}\\ \sigma\to\sigma_{\text{in/out}}}} \exp\left\{i\int\left[\mathcal{L}(x) + \frac{1}{2\alpha}(\partial_\mu A^a_\mu + \alpha m_1 B^a)^a\right]dx\right\}$$

$$\times\prod_x \det M_X\, d\mathcal{A}_\mu d\mathcal{B} d\sigma, \tag{4.45}$$

where the Lagrangian $\mathcal{L}$ is given, as before, by the formula (4.39), and the operator M_X looks as follows:

$$M_X u = (\Box - \alpha m_1^2)u - g\partial_\mu[\mathcal{A}_\mu, u] - \frac{\alpha g m_1}{2}[\mathcal{B}, u] - \frac{\alpha m_1 g}{2}\sigma u. \tag{4.46}$$

As before, $\det M_X$ can be represented by an integral over the fields offictitious particles:

$$\det M_X = \int \exp\left\{i\int \bar{c}^a(x) M_X^{ab} c^b(x) dx\right\}\prod_x d\bar{c}dc. \tag{4.47}$$

The term $m_1 A^a_\mu \partial_\mu B^a$ that is nondiagonal in the fields $\mathcal{A}_\mu, B$ in the Lagrangian (4.38) cancels out with an analogous term in the expression fixing the gauge. As a result, mixed propagators $\mathcal{A}^a_\mu B^b$ are absent. The Feynman rules differ from the ones formulated above in the following points:

1. The propagator corresponding to a vector line is

$$-\delta^{ab}\left[\frac{g_{\mu\nu} - k_\mu k_\nu k^{-2}}{k^2 - m_1^2 + i0} + \frac{k_\mu k_\nu k^{-2}}{(k^2\alpha^{-1} - m_1^2 + i0)}\right]. \tag{4.48}$$

2. The propagator corresponding to the scalar B-line is

$$\frac{\delta^{ab}}{p^2 - m_1^2\alpha + i0}. \tag{4.49}$$

3. The propagator corresponding to fictitious particles is

$$\frac{\delta^{ab}}{p^2 - m_1^2\alpha + i0}. \tag{4.50}$$

4. There appear additional interaction vertices of fictitious particles with the fields B^a and σ. Their explicit forms are easy to derive from the formulas (4.46) and (4.47).

At $\alpha = 0$ these rules obviously coincide with the rules formulated above for the Lorentz gauge.

At this point we conclude the description of examples of interaction of the Yang-Mills field with fields of matter. We hope the examples have been sufficiently typical, and that the reader will be able without difficulty to construct a diagram technique for any arbitrary model either with or without symmetry breaking.

4

Renormalization of Gauge Theories

4.1 Examples of the Simplest Diagrams

The diagram technique developed in the previous chapter allows one to calculate Green functions and probabilities of scattering processes to the acuracy of any order in g. However, direct application of th rules formulated above to the calculation of diagrams containing closed loops leads to a meaningless result—the corresponding integrals diverge at large momenta. Attaching meaning to these expressions is the essence of the renormalization procedure to be studied in the present chapter.

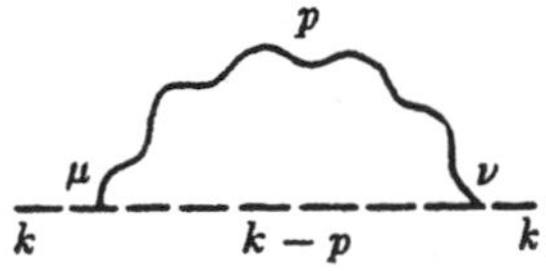

Figure 6. Second-order correction to the Green function of a fictitious particle. The dashed line indicates the propagator of the fictitious particle and the wavy line of the Yang-Mills field.

As the simplest example, we shall consider the second-order correction to the Green function of a fictitious particle in the Yang-Mills theory with the gauge group $SU(2)$. This correction is described in the diagram in Fig. 6. The corresponding analytical expression has the form

$$-\frac{g^2}{k^2+i0}\Sigma(k^2)\frac{1}{k^2+i0}, \tag{1.1}$$

where in the diagonal α-gauge ($\alpha = -1$)

$$\Sigma(k^2) = -\frac{i2k_\nu g_{\mu\nu}}{(2\pi)^4}\int \frac{dp(k-p)_\mu}{(p^2+i0)[(k-p)^2+i0]}. \tag{1.2}$$

As $p \to \infty$ this integral diverges linearly. In order to attach meaning to the integral (1.2), we first introduce an intermediate regularization, replacing the function $(p^2 + i0)^{-1}$ by the regularized expression

$$\frac{1}{(p^2+i0)} \to \frac{1}{(p^2+i0)} - \frac{1}{p^2-\Lambda^2+i0} = \int_0^{\Lambda^2} \frac{d\lambda}{(p^2-\lambda+i0)^2}. \tag{1.3}$$

As $\Lambda \to \infty$ the regularized Green function tends to the initial expression $(p^2+i0)^{-1}$. At finite Λ the integral

$$\Sigma_\Lambda(k^2) = \frac{2ik_\nu}{(2\pi)^4}\int_0^{\Lambda^2} d\lambda \int \frac{dp(k-p)_\nu}{[p^2-\lambda+i0]^2[(k-p)^2+i0]} \tag{1.4}$$

converges. To calculate it, we use the Feynman formula

$$\frac{1}{a^2b} = \int_0^1 \frac{2z\,dz}{[az+b(1-z)]^3}. \tag{1.5}$$

This formula allows us to combine both factors in the denominator of the integral (1.4) into one:

$$\Sigma_\Lambda(k^2) = \frac{2ik_\nu}{(2\pi)^4}\int_0^{\Lambda^2} d\lambda \int_0^1 dz\,2z \times \int \frac{dp(k-p)_\nu}{[(p^2-\lambda+i0)z+(k^2-2pk+p^2+i0)(1-z)]^3}. \tag{1.6}$$

Passing to new variables

$$p \to p + k(1-z), \tag{1.7}$$

we obtain

$$\Sigma_\Lambda(k^2) = \frac{2ik_\nu}{(2\pi)^4}\int_0^{\Lambda^2} d\lambda \int_0^1 dz\,2z \int \frac{dp(kz-p)_\nu}{[p^2+k^2(1-z)z-\lambda z+i0]^3} \tag{1.8}$$

The integral

$$\int dp\, p_\nu f(p^2) \tag{1.9}$$

is equal to zero for reasons of symmetry.

In the remaining integral one can rotate the integration contour through 90° and introduce a new integration variable, $p_o \to ip_o$. As a result, the integral over p takes the form

$$I = -i \int \frac{dp}{(p^2 + c)^3}, \tag{1.10}$$

where integration is performed over the four-dimensional Euclidean space. Calculation of the integral (1.10) gives

$$I = -\frac{i\pi^2}{2c}. \tag{1.11}$$

As result, we obtain the following expression for the function $\Sigma_\Lambda(k^2)$:

$$\Sigma_\Lambda(k^2) = -\frac{k^2}{16\pi^2} \int_0^{\Lambda^2} d\lambda \int_0^1 dz \frac{2z^2}{k^2(1-z)z - \lambda z}. \tag{1.12}$$

The integration over λ is performed explicitly. At $k^2 < 0$ we obtain

$$\Sigma_\Lambda(k^2) = \frac{2k^2}{16\pi^2} \int_0^1 dz \cdot z \ln \frac{k^2(1-z)z - \Lambda^2 z}{k^2(1-z)z}, k^2 < 0. \tag{1.13}$$

As $\Lambda \to \infty$ this expression, as might be expected, diverges logarithmically. The renormalization procedure consists in replacing the integral in (1.13) by an expression obtained by the subtraction from this integral of one or more leading terms of the expansion in a Taylor series (in the case one term is subtracted). Expanding the integrand about the point $k^2 = \mathcal{N}$, we obtain

$$\Sigma_\Lambda(k^2) = \frac{2k^2}{16\pi^2} \left\{ \int_0^1 z\, dz \ln \frac{\mathcal{N}(1-z)z - \Lambda^2 z}{\mathcal{N}(1-z)z} \right.$$
$$\left. + \int_0^1 z\, dz \left[\ln \frac{k^2(1-z)z - \Lambda^2 z}{\mathcal{N}(1-z)z - \Lambda^2 z} - \ln \frac{k^2(1-z)z}{\mathcal{N}(1-z)z} \right] \right\}. \tag{1.14}$$

As $\Lambda \to \infty$ the second and third terms tend to a definite limit equal to

$$\Sigma_R(k^2) = -\frac{k^2}{16\pi^2} \ln \frac{k^2}{\mathcal{N}}. \tag{1.15}$$

The first term has no limit as $\Lambda \to \infty$, and it behaves as

$$-k^2 g^{-2}(1 - \tilde{z}_2) = \frac{k^2}{16\pi^2} \left(\ln \frac{\Lambda^2}{-\mathcal{N}} + \cdots \right). \tag{1.16}$$

The separation (1.14) is not the only one possible. Choosing as the central point of the expansion a point other than $\mathcal{N}$, we would obtain for $\Sigma_R(k^2)$ an expression differing from (1.15) by a finite polynomial in k^2. Thus, the general expression for the renormalized Green function to second order in g^2 has the form

$$-\frac{1}{k^2 + i0} \left(1 + \bar{b}^2 - \frac{g^2}{16\pi^2} \ln \frac{k^2}{\mathcal{N}} \right), \tag{1.17}$$

where b_2 is an arbitrary constant.

Substitution of the renormalized expression (1.15) for the divergent integral (1.2) is equivalent to the redefinition of the original Lagrangian. Indeed, let us replace the Lagrangian for the fictitious particles by the following expression:

$$-\frac{\mathrm{tr}}{2}\bar{c}\partial_\mu \nabla_\mu c \to -\frac{\mathrm{tr}}{2}\{\bar{c}\Box c - g\bar{c}\partial_\mu[\mathcal{A}_\mu, c] + (\bar{z}_2 - 1)\bar{c}\Box c\}, \tag{1.18}$$

where $\bar{z}_2$ is defined by the formula (1.16). Since the last term $\sim g^2$, we shall attribute it to the interaction Lagrangian. Then in the perturbation-theory expansion there will appear, besides the diagram in Fig. 6, a new diagram (Fig. 7), where the cross indicates the vertex responsible for the "counterterm" $(\bar{z}_2 - 1)c\Box c$.

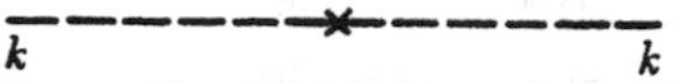

Figure 7.

Obviously the correction to the Green function corresponding to the sum of the diagrams in Figs. 6 and 7 is given by the formula (1.17) (at $\bar{b}_2 = 0$). This simple example shows that the subtraction of the leading terms of the expansion in the Taylor series is equivalent to a change (renormalization) of the original Lagrangian parameters (in this case of the normalization constant of the fictitious-particle wave function).

Let us illustrate this observation by one more example. We shall calculate the third-order correction to the vertex function $\Gamma_{\bar{c}c\mathcal{A}_\mu}$ responsible for the transition of two fictitious particles into one vector particle. The diagrams, contributing to $\Gamma_{\bar{c}c\mathcal{A}_\mu}$ in the third order in g, are presented in Fig. 8.

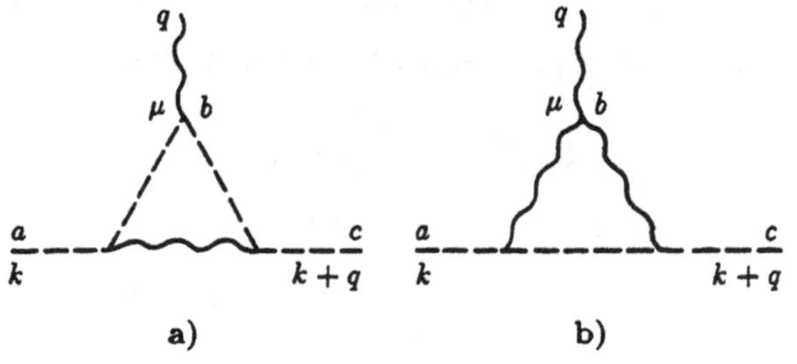

Figure 8. Third-order corrections to the vertex function $\Gamma_{\bar{c}c\mathcal{A}_\mu}$.

For simplicity we shall restrict ourselves to the case of zero momentum transfer $q = 0$.

The integral

$$I_a = -ig^3\varepsilon^{abc}\int \frac{dp(k-p)_\alpha(k-p)_\mu k_\beta g_{\alpha\beta}}{(2\pi)^4(p^2+i0)[(k-p)^2+i0]^2}. \tag{1.19}$$

corresponds to the diagram (a). Introducing intermediate regularization with the formula (1.3) and using the relation

$$\frac{1}{a^2b^2} = \int_0^1 \frac{6z(1-z)dz}{[az+b(1-z)]^4}, \tag{1.20}$$

we write this integral in the form

$$I_a = -\frac{tg^3\varepsilon^{abc}k_\alpha}{(2\pi)^4}\int_0^1 dz \int_0^{\Lambda^2} d\lambda \times \int dp \frac{(k-p)_\alpha(k-p)_\mu 6z(1-z)}{\{[p-k(1-z)]^2+k^2(1-z)z-\lambda z\}^4}. \tag{1.21}$$

The change of variables (1.7) gives

$$I_a = -\frac{ig^3\varepsilon^{abc}k_\alpha}{(2\pi)^4}\int_0^1 d\lambda \int_0^{\Lambda^2} d\lambda \int dp \frac{(kz-p)_\alpha(kz-p)_\mu 6z(1-z)}{\{p^2+k^2(1-z)z-\lambda z\}^4}. \tag{1.22}$$

The odd power of p do not give any contributions for reasons of symmetry. For the same reasons,

$$\int dp p_\mu p_\nu f(p^2) = \frac{1}{4} g_{\mu\nu} \int dp p^2 f(p^2). \tag{1.23}$$

Passing to the Euclidean metric and integrating over p, we obtain

$$I_a = -\frac{ig^3\varepsilon^{abc}k_\alpha}{(2\pi)^4}\left\{\int_0^1 dz \int_0^{\Lambda^2} d\lambda z^2(1-z)k_\alpha k_\mu \frac{\pi^2}{\{k^2(1-z)z-\lambda z\}^2}\right.$$

$$\left. + g_{\alpha\mu}\int_0^1 dz \frac{(-i\pi)^2}{4} z(1-z)\int_0^{\Lambda^2} d\lambda \frac{1}{\{k^2(1-z)z-\lambda z\}}\right\}. \tag{1.24}$$

Integration over λ gives

$$I_a = \frac{g^3}{16\pi^2}\varepsilon^{abc}k^2 k_\mu$$

$$\times \int_0^1 \left\{\frac{1}{k^2(1-z)z-\Lambda^2 z} - \frac{1}{k^2(1-z)z}\right\} z^2(1-z)dz$$

$$+ \frac{g^3}{32\pi^2}\varepsilon^{abc}k_\mu \int_0^1 \ln\frac{k^2(1-z)z-\Lambda^2 z}{k^2(1-x)z}(1-z)dz \quad (k^2<0) \tag{1.25}$$

The first term in the formula (1.25) tends to a definite limit as $\Lambda \to \infty$, and the second one diverges logarithmically. As in the case of the second-order diagram, the expression obtained by subtraction of the first term of the expansion in the Taylor series from the second integral tends to a definite limit. As $\Lambda \to \infty, I_a$ can be expressed as

$$I_a = \frac{g^3\varepsilon^{abc}k_\mu}{64\pi^2}\left(b^1_{\mathcal{N}} - \ln\frac{k^2}{\mathcal{N}}\right) + \frac{g^3\varepsilon^{abc}k_\mu}{64\pi^2}\ln\frac{\Lambda^2}{-\mathcal{N}}, \tag{1.26}$$

where $b^1_{\mathcal{N}}$ is a finite constant, depending on the choice of the point $\mathcal{N}$. The diagram b is calculated in an entirely analogous manner. The corresponding integral has the form

$$I_b = -ig^3\varepsilon^{abc} \times \int \frac{dp}{(2\pi)^4} \frac{(k-p)_\alpha k_\beta \{2p_\mu g_{\sigma\rho} - p_\sigma g_{\mu\rho} - p_\rho g_{\mu\sigma}\} g_{\alpha\sigma} g_{\rho\beta}}{[(k-p)^2 + i0][p^2 + i0]^2}. \tag{1.27}$$

Repeating the computations given above, we obtain

$$I_b = \frac{g^3\varepsilon^{abc}k_\mu}{64\pi^2}\left(b^2_{\mathcal{N}} - 3\ln\frac{k^2}{\mathcal{N}}\right) + \frac{3g^3\varepsilon^{abc}k_\mu}{64\pi^2}\ln\frac{\Lambda^2}{-\mathcal{N}}. \tag{1.28}$$

Subtracting the terms proportional to ln $\Lambda^2/(-\mathcal{N})$ from the sum $I_a + I_b$, we obtain the expression for the renormalized vertex function in the form

$$I^R = \frac{g^3\varepsilon^{abc}k_\mu}{16\pi^2}\left(\tilde{b}_1 - \ln\frac{k^2}{\mathcal{N}}\right), \tag{1.29}$$

where $\tilde{b}_1$ is an arbitrary constant. The subtraction performed is equivalent to the insertion in the Lagrangian of the counterterm

$$\frac{\mathrm{tr}}{2}\{g(\tilde{z}_1 - 1)\bar{c}\partial_\mu[\mathcal{A}_\mu, c]\}, \tag{1.30}$$

where

$$\tilde{z}_1 - 1 = -\frac{g^2}{16\pi^2}\left(\ln\frac{\Lambda^2}{-\mathcal{N}} + \tilde{b}_1\right). \tag{1.31}$$

In conclusion we shall give without computation the expressions for the lowest-order corrections to the Green function of the Yang-Mills field and for the three-point vertex $\Gamma_{\mathcal{A}^3}$, which are described by the diagrams presented in Figs. 9 and 10.

The counterterms which remove the divergences from these diagrams have the form

$$-\frac{\mathrm{tr}}{8}\{(z_2 - 1)(\partial_\nu\mathcal{A}_\mu - \partial_\mu\mathcal{A}_\nu)^2 + 2g(z_1 - 1)(\partial_\nu\mathcal{A}_\mu - \partial_\mu\mathcal{A}_\nu)[\mathcal{A}_\mu, \mathcal{A}_\nu]\}, \tag{1.32}$$

where

$$z_2 - 1 = \frac{g^2 \cdot 5}{24\pi^2}\ln\frac{\Lambda^2}{-\mathcal{N}} + b_2. \tag{1.33}$$

$$z_1 - 1 = \frac{g^2}{12\pi^2}\ln\frac{\Lambda^2}{-\mathcal{N}} + b_1, \tag{1.34}$$

in which b_1 and b_2 are arbitrary finite constants.

As is seen, for the removal of divergences from the diagrams under consideration it is indeed sufficient to redefine the parameters of the original Lagrangian. Note, however, that the counterterms (1.16), (1.30), (1.32) are not, generally speaking, gauge-invariant. Their explicit form depends on the intermediate regularization used and the choice of the subtraction points. In particular, to remove the

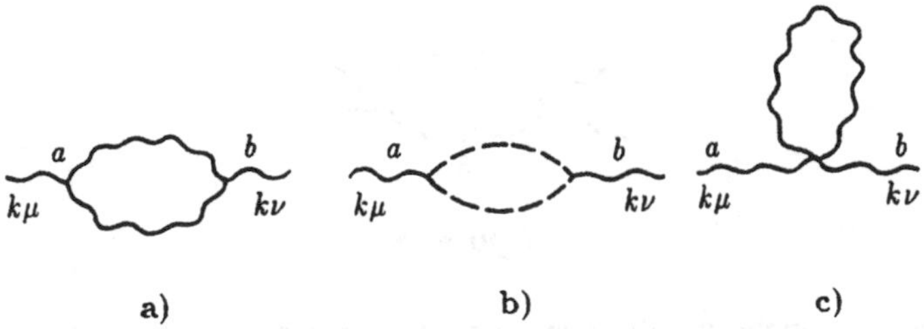

Figure 9. Second-order corrections to the Green function of the Yang-Mills field.

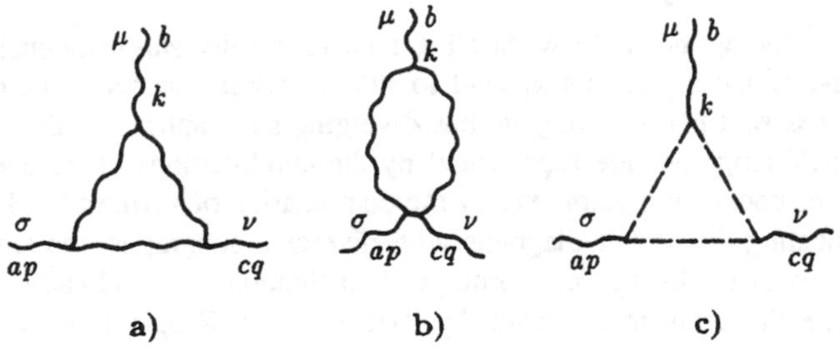

Figure 10. Third-order corrections to the vertex function Γ_{A^3}.

divergences from the two-point Green function it may be necessary to introduce a manifestly noninvariant counterterm

$$\delta m \operatorname{tr}\{\mathcal{A}_\mu \mathcal{A}_\mu\}. \tag{1.35}$$

Of course, it is the renormalized finite matrix elements and not the counterterms themselves that have physical meaning. For the theory to remain self-consistent it is necessary that the renormalized matrix elements satisfy the relativity principle. This requirement entails the specific features of the renormalization procedure in gauge theories, which are to be investigated in the present chapter.

4.2 The R-Operation and Counterterms

In the preceding section we discussed the procedure of removing divergences from the simplest diagrams. The examples considered contain only one integration over dk, and so, when the intermediate regularization is removed, in order to provide for the corresponding functions to tend to a definite limit, it is sufficient to subtract one or more leading terms of their expansion in a Taylor series in the external momenta. As we have seen, such a subtraction is equivalent to the redefinition of the original Lagrangian, that is, to the introduction of counterterms.

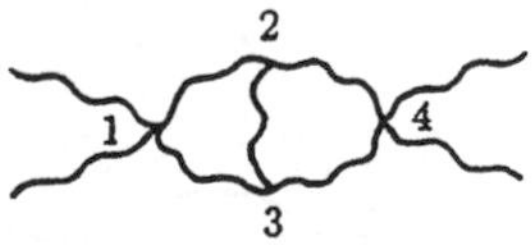

Figure 11.

To more complicated diagrams such as, for example, the one presented in Fig. 11, there correspond integrals of the form

$$\int f(p_1, \ldots, p_m, k_1, \ldots, k_n) dk_1 \ldots dk_n, \tag{2.1}$$

which may diverge not only when all k tend to infinity simultaneously, but also when some of the arguments k_i tend to infinity while the rest remain fixed. In this case it is said that the diagram has diverging subgraphs. For the diagrams in Fig. 11 such subgraphs are represented by the combinations of vertices (1, 2, 3) and the lines connecting them, and by the combination of vertices (2, 3, 4) and the lines connecting them. For diagrams with divergent subgraphs, the simple recipe for the removal of divergences formulated in Section 4.1 is already insufficient. In this case the problem is solved by means of the R-operation of Bogolubov and Parasyuk, which for any Feynman diagram provides a corresponding finite coefficient function. A detailed discussion of the R-operation may be found in the book of N. N. Bogolubov and D. V. Shirkov, and we shall not repeat it here. For our aims it is sufficient to know that the R-operation is equivalent to the insertion in the Lagrangian of counterterms, which may be represented as series in the coupling constant. In order to formulate the corresponding recipe we shall need several definitions. A diagram is said to be connected if it cannot be separated into parts that are not connected to each other by lines. A diagram is called strongly connected or one-particle-irreducible if it cannot be transformed into disconnected diagrams by the removal of a single line. A strongly connected diagram with all external lines cut off will be called a proper vertex function and denoted as $\Gamma(x_1, \ldots, x_n)$. A strongly connected Green function $G(x_1, \ldots, x_n)$ is expressed in terms of the proper vertex function $\Gamma(x_1, \ldots, x_n)$ by the relation

$$G(x_1, \ldots, x_n) = \int dx'_1 \ldots dx'_n G_1(x_1 - x'_1) \ldots G_n(x_n - x'_n) \Gamma(x'_1, \ldots, x'_n), \tag{2.2}$$

where $G_i(x_i - x'_i)$ is the two-particle Green function corresponding to the i-th external line. The topological structure of the diagrams is conveniently characterized by the number of independent cycles contained in the given diagram. Diagrams with one cycle are called one-loop diagrams; with two cycles, two-loop diagrams; etc. Diagrams with a given number of loops are terms of the same order in the quasiclassical expansion in Planck's constant $\hbar$ of the S-matrix or of the generating functional for the Green functions. For this reason they form an invariant combination; that is, all the symmetry properties of the complete S-matrix are satisfied independently for a combination of diagrams with a fixed number of loops.

To characterize the procedure for the removal of divergences, the notion of a diagram index is introduced. Let the coefficient function with the Fourier transform

$$J(k) = \int \prod_{1\le q\le n} \delta\left(\Sigma p - k_q\right) \prod_{l=1}^{L} D_l(p_l) dp_l, \tag{2.3}$$

where index q numbers vertices and l numbers the internal lines, correspond to a strongly connected diagram. In the argument of the δ-function there is the algebraic sum of momenta that enter into the vertex number q. The Green function $D_1(p_1)$ has the form

$$D_l(p_l) = Z\left(p_l\right)\left(m_l^2 - p_l^2\right)^{-1}, \tag{2.4}$$

where $Z(p_1)$ is a polynomial of degree r_1.

Let us perform a scalar transformation of all the momentum variables (and masses): $p_i, k_i \to ap_i, ak_i$. If the integral $J(k)$ converges, then under such a transformation it will be multiplied by a^ω, where the diagram index ω consists of the following factors: Each internal line contributes r_i-2, resulting in $\sum_{l=1}^{L}(f_1-2)$, where L is the total number of internal lines. In the formula (2.3) integration is performed over L variables p_1; however, $n-1$ integrals are removed by δ-functions (one δ-function expresses the conservation law for the total momentum). Therefore there remain $4(L-n+1)$ independent differentials, which give a total contribution of $4(L-n+1)$. If the interaction Lagrangian contains derivatives, then each vertex with m derivatives introduces an additional factor m. Summing these factors, we obtain

$$\omega = \sum_l (r_l + 2) - 4(n-1) + mn. \tag{2.5}$$

The index ω defines the degree of growth of the coefficient function when a uniform extension of all the momentum variables takes place. At $\omega \geq 0$ this definition, in general, loses its sense, since the corresponding integral diverges. In this case the diagram index defines the superficial degree of growth. From the nonnegativeness of the diagram index follows the divergence of the corresponding integral. The opposite is not true in general, since the index of a diagram characterizes its behavior only when a simultaneous extension of all momenta occurs, and has nothing to do with its behavior when some of the integration variables tend to infinity while the rest remain fixed. In other words, a diagram with a negative index may have divergent subgraphs. Negativeness of the index is a sufficient condition for the convergence of primitively divergent diagrams, that is, of diagrams that become convergent if any internal line is broken. The diagrams considered in the first section are primitively divergent, whereas the diagram in Fig. 11 has divergent subgraphs (1, 2, 3) and (2, 3, 4). Obviously, one-loop diagrams can be only primitively divergent.

We can now formulate a recipe for the removal of divergences from arbitrary diagrams by the insertion of counterterms. First of all we shall introduce an intermediate regularization, making all the integrals convergent, for instance, by means of the formula (1.3).

First, let us consider one-loop diagrams, As we have already seen, for the corresponding coefficient functions to tend to a definite limite when the regularization is removed, it is sufficient to subtract from them several leading terms of the expansion in the Taylor series in the external momenta. Such a subtraction is, in turn, equivalent to the insertion in the Lagrangian of counterterms, that is, to the substitution of $\mathcal{L}_\Lambda + \Delta\mathcal{L}_1$ for the original regularized Lagrangian $\mathcal{L}_\Lambda$.

The explicit expression for the counterterms $\Delta\mathcal{L}_1$ is constructed in the following way. Let G_s^n be a strongly connected diagram with n vertices and s external lines $\mathcal{A}_\mu$, having a nonnegative index ω. To it there corresponds the proper vertex function $\Gamma_s^n(x_1, \ldots, x_n)$. The subtracted polynomial consists of the first terms in the expansion of the Fourier transform of Γ_s^n in the Taylor series, and in the coordinate representation it has the form

$$Z_{(\mu)}\left(\frac{\partial}{\partial x_i}\right)\delta(x_1 - x_2)\ldots\delta(x_{s-1} - x_s), \tag{2.6}$$

where Z is a symmetric polynomial of order ω. In order to obtain the counterterm corresponding to the given diagram, it is necessary to multiply the expression (2.6) by the product

$$\mathcal{A}_{\mu_1}(x_1)\ldots\mathcal{A}_{\mu_s}(x_s), \tag{2.7}$$

and then to sum the obtained expression over $\mu_1, \ldots, \mu_s$ and integrate over all variables $x_1, \ldots, x_n$ except one. (If besides vector external lines, the diagram contains others—for instance, spinor and scalar lines—then all the reasoning remains the same except that symmetrization is applied only to lines of the same type.)

We shall now construct two-loop diagrams using as a Langrangian $\mathcal{L}_\Lambda + \Delta\mathcal{L}_1$. The two-loop diagrams thus constructed do not now contain divergent subgraphs; that is, when the intermediate regularization is removed, divergence appears only with the simultaneous tending of all integration variables to infinity. This fact is quite obvious if the divergent subgfraphs do not overlap, as happens, for example, with the diagram presented in Fig. 12. In this case, the counterterm $\Delta\mathcal{L}_1$, when we remove the divergence from the subgraph (2.3), has the form

$$-\frac{(z_2 - 1)}{4}(\partial_\mu A_\nu^a - \partial_\nu A_\mu^a)^2, \tag{2.8}$$

and the Lagrangian $\mathcal{L} + \Delta\mathcal{L}_1$ generates, in addition to the diagram in Fig. 12, also the one presented in Fig. 13, where the cross indicates the vertex (2.8). The integral corresponding to the sum of the diagrams in Figs. 12 and 13 diverges only when all momenta tend simultaneously to infinity, and for the removal of the divergence it is again sufficient to subtract from the integral the first two terms of the expansion in the Taylor series, which is equivalent to the insertion in the Lagrangian of a new counterterm $\Delta\mathcal{L}_2$:

$$\mathcal{L} \rightarrow \mathcal{L} + \Delta\mathcal{L}_1 + \Delta\mathcal{L}_2. \tag{2.9}$$

Figure 12.

Figure 13.

The proof of an analogous statement when overlapping divergent subgraphs are present (for instance, as in Fig. 11) is more complicated, and we shall not present it here.

Proceeding in this manner, we come to the renormalized Lagrangian

$$\mathcal{L}_R = \mathcal{L} + \Delta\mathcal{L}_1 + \ldots + \Delta\mathcal{L}_n, \tag{2.10}$$

where $\Delta\mathcal{L}_i$ are local polynomials in the fields and their derivatives, for which all diagrams containing not more than n loops converge to a finite limit when the intermediate regularization is removed. Obviously, in the framework of perturbation theory, we may thus calculate the Green functions to any finite order n. With increasing n the total number of counterterms of various types may turn out to remain finite. (We call the functional dependence of a counterterm on the fields the type of the counterterm) In this case it is said that the theory is normalized. A renormalizable theory is determined by a finite number of parameters, having the meaning of physical charges and masses. But if the number of the counterterm types increases infinitely (that is, in the higher orders of perturbation theory there appear structures containing more and more fields and their derivatives), then the theory is called nonrenormalizable. Since the insertion of a new counterterm is equivalent to the appearance of a new arbitrary constant (the position of the point of subtraction), nonrenormalizable theories are not determined by a finite number of parameters. For nonremormalizable Lagrangians the perturbation-theory method seems to be useless, and we shall not consider them here.

The explicit form of the counterterms depends on the concrete intermediate regularization and on the choice of the subtraction point, that is, the center of expansion in the Taylor series. An inconvenient choice of regularization may render the analysis of the renormalized theory extremely difficult. In the case of gauge theories the so-called invariant regularizations, conserving the formal symmetry properties of the nonrenormalized theory, are especially convenient.

4.3 Invariant Regularizations. The Pauli-Villars Procedure

The counterterm form of the R-operation is convenient for the investigatin of Yang-Mills fields, since it allows us to take into account symmetry properties in a simple and explicit manner. As we have already seen in the previous chapter, the relativity principle allows us to construct a perturbation theory for Yang-Mills fields, proceeding from various gauges. The gauges in which the S-matrix is formally unitary (the Coulomb or Hamilton gauges for the massless Yang-Mills field, the unitary gauge for the theory with spontaneously broken symmetry) are inconvenient from the point of view of the renormalization procedure. In the first two cases there is no manifest relativistic invariance, and in the latter case explicit renormalizability is absent. Significantly more convenient in this sense are the manifestly covariant gauges, such as the Lorentz one, for which, as we shall see, renormalizability is obvious. However, in the Lorentz gauge we cannot construct a Hamiltonian formulation of the theory, and therefore the unitarity of the S-matrix is not obvious. From the viewpoint of the operator formalism, the S-matrix in the Lorentz gauge acts in the "big" space containing both physical and nonphysical states (longitudinal and time "photons," scalar fermions, Goldstone bosons) and, generally speaking, is unitary only in this space, in which the metric is indefinite. The unitarity of the S-matrix in the physical subspace, the states of which correspond to fields of matter and to transverse vector quanta, is a consequence of the relativity principle, according to which all observables are independent of the gauge condition actually chosen. This is confirmed by the explicit calculations of the previous chapter, where it was shown that an explicitly unitary generating functional for the coefficient functions of the S-matrix in the Coulomb gauge may be transformed identically into a functional corresponding to the Lorentz gauge. The reasoning given, however, was of a formal character, since we did not pay attention to the divergences appearing in the perturbation-theory calculations of these functionals. Indeed, in quantum theory the relativity principle should be applied to renormalized entities free of divergences. The transfer of this principle to a renormalized theory is not trivial. Renormalization is equivalent to the redefinition of the original Lagrangian. Therefore it is necessary to prove that the renormalized Lagrangian is gauge-invariant. Then we may apply it to the reasoning of the preceding chapter and rigorously prove the equivalence of various gauges and, consequently, the unitarity of the S-matrix.

The previous statement needs to be clarified. As we have already seen, the explicit form of the renormalized Lagrangian depends on the intermediate regularization used. What we have said above applies only to the invariant intermediate regularization, that is, to the regularization that conserves the formal symmetry properties of the nonrenormalized theory. This, of course, does not mean that for the calculation of the S-matrix one is not to use a noninvariant regularization. In that case, however, the regularized theory is gauge-noninvariant; and the reasoning of the previous chapter, which demonstrates the equivalence of various gauges, is not applicable to it. The relativity principle is now valid only for the renormalized S-matrix, when the regularization is removed. All this complicates the proof of unitarity and makes it less clear. Therefore we shall start describing the renormalization procedure for the Yang-Mills theory by constructing a gauge-invariant intermediate regularization. The specific features of the invariant regularization of gauge theories are due to the interaction of Yang-Mills fields in vacuum. The interaction with fields of matter does not introduce any difficulties: The corresponding diagrams are regularized by means of the obvious generalization of the gauge-invariant Pauli-Villars procedure.

We shall show this by using the example of the interaction of the Yang-Mills field with a spinor field ψ described by the Lagrangian (1.3.1) of Chapter 1. The generating functional for the Green functions has the form

$$\begin{aligned} Z(J_\mu, \bar{\eta}, \eta) = N^{-1} \int \exp\Big\{ i \Big[\mathcal{L}_{YM} + \frac{1}{2\alpha}(\partial_\mu A_\mu^a)^2 \\ + i\bar{\psi}(\hat{\partial} - g\Gamma^a \hat{A}^a)\psi - \mu_0 \bar{\psi}\psi + J_\mu^a A_\mu^a \\ + \bar{\eta}\psi + \bar{\psi}\eta \Big] dx \Big\} \det M \prod_x dA_\mu d\bar{\psi} d\psi. \end{aligned} \tag{3.1}$$

Divergent diagrams not containing internal vector lines (spinor cycles) are regularized in the same manner as in electrodynamics, that is, by subtracting analogous cycles along which the spinor fields with masses μ_i propagate. Actually, if we are interested only in spinor cycles, we may set the sources $\bar{\eta}, \eta$ equal to zero. The remaining Gaussian integral over ψ and $\bar{\psi}$ is calculated explicitly. It is equal to $\det X_0$, where

$$X_0 = i\gamma_\mu \partial_\mu - \mu_0 - ig\Gamma^a \gamma_\mu A_\mu^a. \tag{3.2}$$

The regularization consists in the substitution for $\det X_0$ of the product

$$\det X_0 \to \det X_0 \prod_{j=1}^n (\det X_j)^c j = \exp\{\operatorname{Tr}\ln X_0 + \sum_{j=1}^n c_j \operatorname{Tr}\ln X_j\}, \tag{3.3}$$

where the operators X_j are constructed analogously to the operator X_0:

$$X_j = i\hat{\partial} - \mu_j - ig\Gamma^a \hat{A}^a, \tag{3.4}$$

and the coefficients c_j satisfy the conditions

$$\Sigma c_j + 1 = 0, \qquad \Sigma c_j \mu_j^2 = 0. \tag{3.5}$$

In order to verify that the substitution (3.3) really regularizes the spinor cycles, let us represent $\det X_j$ in the form

$$\det X_j = \det(i\hat{\partial} - \mu_j)\det\{1 - g(i\hat{\partial} - \mu_j)^{-1} i\Gamma^a \hat{A}^a\}. \tag{3.6}$$

The first factor does not depend on the fields A_μ and therefore can be included in the normalization constant N. The second factor may be transformed into

$$\exp\{\operatorname{Tr}\ln\{1 - g(i\hat{\partial} - \mu_j)^{-1} i\Gamma^a \hat{A}^a\}\}$$

$$= \exp\Big\{ -\Big[\frac{(ig)^2}{2} \operatorname{tr} \int [\Gamma^a \hat{A}^a(x_1) S^j(x_1 - x_2) \times \Gamma^{a_2} \hat{A}^{a_2}(x_2) S^j(x_2 - x_1)] dx_1 dx_2 + \dots$$

$$+ \frac{(ig)^n}{n} \operatorname{tr} \int \Gamma^{a_1} \hat{A}^{a_1}(x_1) S^j(x_1 - x_2) \dots \Gamma^{a_n} \hat{A}^{a_n}(x_n) S^j(x_n - x_1) dx_1 \dots dx_n \Big]\Big\}, \tag{3.7}$$

where $S^j(x)$ is the spinor Green function

$$S^j(x) \equiv (i\hat{\partial} - \mu_j)^{-1} = \frac{-1}{(2\pi)^4} \int \frac{\mu_j + \hat{p}}{\mu_j^2 - p^2 - i0} e^{-ipx} dp. \tag{3.8}$$

Passing to Fourier transforms, the n-th term in the exponential can be written as

$$\text{const} \cdot \int \Big[\int dp \frac{\operatorname{tr}[\gamma_{\nu_1}(\mu_j + \hat{p}) \dots \gamma_{\nu_n}(\mu_j + \hat{p} + \hat{k}_{n-1})]}{(\mu_j^2 - p^2)(\mu_j^2 - (p + k_1)^2) \dots (\mu_j - (p + k_{n-1})^2)}$$

$$\times \operatorname{tr}\Big[\Gamma^{a_1} A^{a_1}_{\nu_1}(k_1) \dots \Gamma^{a_n} A^{a_n}_{\nu_n}(k_n)\Big] \times \delta(k_n - k_1 - \dots - k_{n-1})\Big] dk_1 \dots dk_n. \tag{3.9}$$

Here the first trace is related to the spinor indices and the second one to the internal degrees of freedom. At $n \leq 4$ the integral over p diverges. At large p the integrand in this integral may be represented as a series in μ_i,

$$\frac{p_n(p) + \mu_j^2 p_{n-2}(p) + \dots + \mu_j^n}{P_{2n}(p) + \mu_j^2 P_{2n-2}(p) + \dots + \mu_j^{2n}} = \frac{P_n(p)}{P_{2n}(p)} + \frac{P_n(p)}{P_{2n}(p)}$$

$$\times \left[\frac{P_{n-2}(p)}{P_n(p)} - \frac{P_{2n-2}(p)}{P_{2n}(p)} \right] \mu_j^2 + \dots, \tag{3.10}$$

where $P_j(p)$ is a polynomial of the order j in p. The coefficient of μ_j^{2k} at large p behaves as p^{-n-2k}. If the coefficients c_j satisfy the conditions (3.5), then the two highest-order terms in the asymptotic expansion in p of the integrand in the sum (3.3) fall out, and the aysmptotic behavior of the regularized expression is p^{-n-4}. Thus, all the integrals over p converge.

The regularized generating functional can be represented by a path integral in which the exponent contains the local action, if one uses the representation of $\det X_j$ in the form

$$[\det X_J]^{\pm 1} = \int \exp\Big\{ i \int [i\bar{\psi}_j(\hat{\partial} - g\Gamma^a \hat{A}^a)\psi_j - \mu_j \bar{\psi}_j \psi_j] dx \Big\} \prod_x d\bar{\psi}_j d\psi_j, \tag{3.11}$$

where $\bar{\psi}_i, \psi_j$ are the auxiliary spinor variables. The exponent of the determinant on the left-hand side of this equality depends on the commutation properties of the fields ψ_j. The exponent $+1$ corresponds to anticommuting variables, and -1 corresponds to commuting variables. By choosing integers for the coefficients c_j in the formula (3.3), we can represent the regularizing factor $\prod_{j=1}^{n}(\det X_j)^{c_j}$ in the form

$$\prod_{j=1}^{n}(\det X_j)^{c_j} = \int \exp\left\{ i \int \sum_{j=1}^{n} \left[\sum_{k=1}^{|c_j|} [i\bar{\psi}_{jk}(\hat{\partial} - g\Gamma^a \hat{A}^a)\psi_{jk} \right. \right.$$

$$\left. \left. -\mu_j \bar{\psi}_{jk}\psi_{jk}] \right] dx \prod_{x,j,k} d\bar{\psi}_{jk} d\psi_{jk} \right\}. \tag{3.12}$$

Here the coefficients c_j and the masses μ_j are assumed to satisfy the conditions (3.5). The commuting auxiliary fields $\bar{\psi}_{jk}, \psi_{jk}$ correspond to the negative coefficients c_j, and the anticommuting auxiliary fields correspond to the positive coefficients c_j.

Adding to the action in the exponent (3.1) the action from the right-hand side of the equality (3.12), we obtain the regularized generating functional in the form of a path integral of $\exp\{i \times \text{local action}\}$. The action (3.12) is manifestly invariant with respect to simultaneous gauge transformations of the fields $A_\mu, \bar{\psi}_{jk}, \bar{\psi}_{jk}$, and therefore the regularization (3.3) does not change the symmetry properties of the generating functional.

The generalization of this procedure to the case when the Yang-Mills field interacts also with a scalar field is obvious. The only difference is in that, since scalar fields are commuting entities, the sum of the closed cycles is equal to $(\det Y_0)^{-1}$ and the regularization consists in the substitution

$$(\det Y_0)^{-1} \to (\det Y_0)^{-1} \prod_j (\det Y_j)^{-c_j}. \tag{3.13}$$

The Pauli-Villars regularization is applicable in those cases when the interaction Lagrangian is quadratic in the fields forming divergent cycles. Therefore, it cannot be generalized to the Yang-Mills field itself. Here one has to resort to more sophisticated methods. We shall further restrict ourselves to the consideration of the Yang-Mills field in vacuum.

At present there exist two methods for invariant regularization of non-Abelian gauge theories: the method of higher covariant derivatives and the method of dimensional regularization.

The first method is, actually, an invariant generalization of the standard regularization procedure, when free propagators are regularized by the subtraction

$$-\frac{1}{k^2} \to -\frac{1}{k^2} - \frac{1}{\Lambda^2 - k^2} = -\frac{1}{k^2 - \Lambda^{-2}k^4} \tag{3.14}$$

(for simplicity the scalar propagator is written). Such a subtraction is equivalent to the insertion in the Lagrangian of terms with higher derivatives:

$$\frac{1}{2}\partial_\mu\varphi\partial_\mu\varphi \to \frac{1}{2}\partial_\mu\varphi\partial_\mu\varphi + \frac{1}{2\Lambda^2}\Box\varphi\Box\varphi. \tag{3.15}$$

In case of the Yang-Mills field such a procedure violates the gauge invariance, since an ordinary derivative is not a covariant object. A natural generalization of the regularization (3.15) consists in adding to the Yang-Mills Lagrangian a term containing higher covariant derivatives, for example,

$$\mathcal{L}_{YM} \to \mathcal{L}_{YM}^{\Lambda} = \frac{1}{8}\mathrm{tr}\left\{\mathcal{F}_{\mu\nu}\mathcal{F}_{\mu\nu} + \frac{1}{\Lambda^2}\nabla_\alpha\mathcal{F}_{\mu\nu}\nabla_\alpha\mathcal{F}_{\mu\nu}\right\}$$

$$= \frac{1}{8}\mathrm{tr}\left\{\mathcal{F}_{\mu\nu}\mathcal{F}_{\mu\nu} + \frac{1}{\Lambda^2}(\partial_\alpha\mathcal{F}_{\mu\nu} - g[\mathcal{A}_\alpha, \mathcal{F}_{\mu\nu}])^2\right\}. \quad (3.16)$$

The substitution (3.16) leads to the desired modification of the free propagator. However, the cost of achieving invariance is the appearance of new vertices in the interaction Lagrangian. Below we shall discuss this regularization in more detail, but now we just point out that due to the appearance of new vertices with derivatives, the regularization is only partial—in the regularized theory the second-, third-, and fourth-order diagrams remain divergent. Thus the method of higher derivatives alone does not solve the problem completely, but only reduces the problem to the investigation of a supernormalizable theory, that is, a theory generating a finite number of divergent diagrams. Below it will be shown that the remaining diagrams can be regularized by means of a somewhat modified Pauli-Villars procedure. As a result we shall describe an explicitly invariant Lagrangian that generates convergent (for finite regularization parameters) Feynman diagrams. The defect of this method is that it is relatively cumbersome. Due to the appearance of new vertices in the interaction Lagrangian, the number of diagrams is greatly increased, impeding practical calculations. However, for investigating fundamental problems of unitarity and renormalization this method is the most convenient one, because the existence of a manifestly invariant expression for the regularized action allows us automatically to apply to the regularized case the reasoning of the previous chapter on the equivalence of various gauges, and thus to prove the unitarity of the renormalized theory.

Unlike the method of higher covariant derivatives, dimensional regularization is not reduced to some modification of the original Lagrangian, but deals directly with the Feynman diagrams. This method is based on two observations:

1. The formal symmetry relations between Green functions (generalized Ward identities) do not depend on the dimensionality of the space-time (n).

2. At sufficiently small or complex n all diagrams correspond to convergent integrals.

Thus, generalized Ward identities can be proven rigorously in the region of n where all integrals converge, and then by analytical continuation one can pass over to $n = 4$.

The method of dimensional regularization has turned out to be convenient for the calculation of concrete diagrams and is quite widely used in practical calculations. It has, however, some shortcomings from the viewpoint of the investigation of matters of principle.

Since if n is noninteger or complex, no Lagrangian can be found to correspond to the regularized theory, the simple proof of unitarity based on the change of variables in the path integral is not applicable, and it is then necessary to deal directly with Feynman diagrams, which is significantly more laborious. Additional difficulties arise for the regularization of theories containing fermions. Since the algebra of the γ-matrices depends crucially on the dimensionality of the space, such theories require special consideration.

Thus, the method of higher covariant derivatives and the method of dimensional regularization, in a sense, complement each other—the first method is more convenient for general proofs that require, in fact, only the existence of the invariant regularized action, whereas the second method is more effective for calculating concrete processes.

4.4 The Method of Higher Covariant Derivatives

The regularization will include two steps: First, by inserting in the Langrangian higher covariant derivatives we shall pass to the super-renormalizable theory, in which only a finite number of one-loop diagrams are involved; and then we shall regularize the one-loop diagrams, using the modified Pauli-Villars procedure.

The modification of the Lagrangian (3.16) is insufficient to provide for the convergence of all the diagrams containing more than one loop. The Lagrangian (3.16), although corresponding to the super-renormalizable theory, generates a divergent two-loop self-energy diagram of the fourth order. In order to remove this divergence also, we insert in the Lagrangian a term containing fourth-order covariant derivatives:

$$\mathcal{L}_{YM} \rightarrow \mathcal{L}_{\Lambda} = \frac{1}{8}\mathrm{tr}\left\{\mathcal{F}_{\mu\nu}\mathcal{F}_{\mu\nu} + \frac{1}{\Lambda^4}\nabla^2\mathcal{F}_{\mu\nu}\nabla^2\mathcal{F}_{\mu\nu}\right\}. \tag{4.1}$$

The gauge invariance of the regularized Lagrngian is obvious.

The regularized generating functional for the Green functions has the form

$$Z_\Lambda(J_\mu) = N^{-1}\int \exp\left\{i\int\left[\mathcal{L}_\Lambda(x) + \frac{1}{2\alpha}\{f(\Box)\partial_\mu A^a_\mu\}^2 + J^a_\mu A^a_\mu\right]dx\right\}\det M\prod_x dA, \tag{4.2}$$

where $f(\Box)$ is an arbitrary function of the d'Alambertian operator, defining the concrete form of the generalized α-gauge. The regularized free propagator of the Yang-Mills field is constructed in the usual manner:

$$D^{ab}_{\mu\nu} = \delta^{ab}\left[-\left(g_{\mu\nu} - \frac{k_\mu k_\nu}{k^2}\right)\frac{1}{k^2+\Lambda^{-1}k^6} - \frac{\alpha k_\mu k_\nu}{k^4 f^2(-k^2)}\right]. \tag{4.3}$$

In the Lorentz gauge ($\alpha = 0$) at large k the propagator behaves as k^{-6}. If $\alpha \neq 0$, then we shall choose the function $f(-k^2)$ to be such as not to impair the asymptotic behavior of the propagator as $k \to \infty$ and $k \to 0$, for example,

$$f(-k^2) = k^2 - \mathcal{N}^2, \tag{4.4}$$

where $\mathcal{N}^2$ is an arbitrary parameter.

Explicitly writing out the term $\sim \Lambda^{-4}$ in the Lagrangian (4.1),

$$\frac{\mathrm{tr}}{8\Lambda^4}\{[\Box(\partial_\nu \mathcal{A}_\mu - \partial_\mu \mathcal{A}_\nu)]^2 - 2[\Box(\partial_\nu \mathcal{A}_\mu - \partial_\mu \mathcal{A}_\nu)]\partial_\alpha[\mathcal{A}_\alpha, \mathcal{F}_{\mu\nu}] + \dots$$
$$+ [\mathcal{A}_\beta[\mathcal{A}_\alpha[\mathcal{A}_\mu, \mathcal{A}_\nu]]]\}^2, \tag{4.5}$$

we see that it generates vertices with three, four, five, six, seven, and eight outgoing lines. The maximal number of derivatives in each of these vertices is 5, 4, 3, 2, 1, 0, respectively. Let us now calculate the index for an arbitrary diagram. Taking into account that in our case $r_1 = 4$, we obtain that the index of a diagram, containing n_k vertices with k outgoing lines L_{in} and internal lines is given by the formula

$$\omega \leq 4 + n_3 - n_5 - 2n_6 - 3n_7 - 4n_8 - 2L_{\text{in}}$$
$$= 6 - 2\prod - n_3 - 2n_4 - 3n_5 - 4n_8 - 5n_7 - 6n_8, \tag{4.6}$$

where Π is the number of closed loops.

It is not difficult to see that only integrals corresponding to one-loop second-order diagrams with two external lines, or third-order diagrams with three external lines, or fourth-order diagrams with four external lines, can be divergent. Analogous divergent diagrams are also generated by the determinant det M. Converging integrals correspond to all other diagrams, including one-loop diagrams with external lines corresponding to fictitious c-particles.

We shall, first of all, show that for the removal of divergences from the one-loop diagrams it is sufficient to introduce gauge invariant counterterms.

The total contribution of the closed cycles with external lines $\mathcal{A}_\mu$ may be represented as

$$Z_0 = N^{-1} \int \exp\left\{ i \left[\frac{1}{2} \frac{\delta^2 S_\Lambda}{\delta \mathcal{A}_\mu^a(x) \delta \mathcal{A}_\nu^b(y)} q_\mu^a(x) q_\nu^b(y) \right] dx\, dy \right.$$
$$\left. + \frac{1}{2\alpha} \int \{ f(\Box) \partial_\mu q_\mu^a(x) \}^2 dx \right\} \det M(\mathcal{A}) \prod_x dq_\mu. \tag{4.7}$$

Unlike the generating functional for one-loop diagrams of matter fields, considered in the preceding section, the functional $Z_0(\mathcal{A}_\mu)$ has no such property. This is due to the presence in it of a term fixing the gauge, and of cycles of fictitious particles, which violate the explicit gauge invariance. Nevertheless, as we shall now show, the divergent part of the functional $Z_0(\mathcal{A})$ is invariant under gauge transformations of the fields $\mathcal{A}_\mu$; to an accuracy of finite terms, which do not need regularization, the functional (4.7) may be transformed into the manifestly invariant form

$$Z_0 = N^{-1} \int \exp\left\{ i \int \left[\frac{1}{2} \frac{\delta^2 S_\Lambda}{\delta \mathcal{A}_\mu^a(x) \delta \mathcal{A}_\nu^b(y)} q_\mu^a(x) q_\nu^b(y) \right] dx\, dy \right\}$$
$$\times \det \nabla_\mu^2 \prod_x \delta(\nabla_\mu q_\mu) dq_\mu + \ldots, \tag{4.8}$$

where

$$\nabla_\mu q_\mu = \partial_\mu q_\mu - g[\mathcal{A}_\mu, q_\mu], \tag{4.9}$$

and the dots denote the finite terms not to be regularized.

Invariance of the functional (4.8) follows from the derivatives $\frac{\delta^2 S_\Lambda}{\delta \mathcal{A}_\mu^a(x) \cdot \delta \mathcal{A}_\nu^b(y)}$ being transformed contragradiently under gauge transformations of the fields $\mathcal{A}_\mu$. From the invariance of S follows

$$\frac{\delta^2 S_\Lambda}{\delta \mathcal{A}_\mu^a(x) \delta \mathcal{A}_\nu^b(y)}$$
$$= \frac{\delta^2 S_\Lambda}{\delta \mathcal{A}'^c_\mu(x) \delta \mathcal{A}'^d_\nu(y)} \cdot (\delta^{ac} \delta^{b_l} - g t^{cal} u^l(x) \delta^{bd} - g t^{bdl} u^l(y) \delta^{ac} + \ldots), \tag{4.10}$$

Therefore, if together with the gauge transformation $\mathcal{A}_\mu \to \mathcal{A}_\mu^\omega$ one performs the compensating change of integration variables

$$q_\mu \to \omega q_\mu \omega^{-1} \tag{4.11}$$

the integral (4.8) will remain unchanged. (The factors $\det \nabla_\mu^2$ and $\delta(\nabla_\mu q_\mu)$ are evidently invariant with respect to that change.)

To demonstrate the equivalence between the expressions (4.7) and (4.8), we shall use a device with which we are already familiary. We multiply the functional (4.8) by "unity,"

$$1 = \Delta_W(\mathcal{A}_\mu) \int \prod_x \delta(\partial_\mu(q_\mu + \nabla_\mu u) - W(x)) du, \tag{4.12}$$

and change the variables,

$$q_\mu \to q_\mu - \nabla_\mu u. \tag{4.13}$$

As a result, the functional (4.8) assumes the form

$$Z_0 = N^{-1} \int \exp\left\{ i \int \left[\frac{1}{2} \frac{\delta^2 S_\Lambda}{\delta \mathcal{A}_\mu^a(x) \delta \mathcal{A}_\nu^b(y)} [q_\mu - \nabla_\mu u]^a \right.\right.$$
$$\left.\left. \times [q_\nu - \nabla_\nu u]^b \right] dx\, dy \right\} \det \nabla_\mu^2 \prod_x \delta(\nabla_\mu q_\mu - \nabla^2 u)$$
$$\times \delta(\partial_\mu q_\mu - W(x)) \Delta_W(\mathcal{A}_\mu) dq_\mu\, du. \tag{4.14}$$

Integration over u is removed by the δ-function. The Jacobian that then arises cancels out with $\det \nabla_\mu^2$. Noting that on the surface $\partial_\mu q_\mu = W$ the functional $\Delta_W(\mathcal{A}_\mu)$ equals $\det M$, we rewrite (4.14) in the form

$$Z_- = N^{-1} \int \exp\left\{ i \int \left[\frac{1}{2} \frac{\delta^2 S_\Lambda}{\delta A_\mu^a(x) \delta A_\nu^b(y)} [q_\mu - \nabla_\mu \nabla^{-2} \nabla_\rho q_\rho]^a \right.\right.$$

$$\left.\left. \times [q^\nu - \nabla_\nu \nabla^{-2} \nabla_\rho q_\rho]^b \right] dx\, dy \right\} \prod_x \delta(\partial_\mu q_\mu - W(x)) \det M dq_\mu. \quad (4.15)$$

Integrating (4.15) with the weight $\exp\{\frac{1}{2\alpha} \int \{f(\Box)W\}^2 dx\}$ we obtain a functional differing from (4.7) only by the substitution of

$$q_\mu - \nabla_\mu \nabla^{-2} \nabla_\rho q_\rho. \quad (4.16)$$

for q_μ in the exponent.

We shall show that the resulting additional diagrams correspond to converging integrals. To this end we shall take advantage of the relation

$$\int \frac{\delta^2 S_\Lambda}{\delta A_\mu^a(x) \delta A_\nu^b(y)} [\nabla_\mu \varphi(x)]^a dx = g t^{abc} \frac{\delta S_\Lambda}{\delta A_\nu^a(y)} \varphi^c(y), \quad (4.17)$$

following from gauge invariance of the action S_Λ. This relation permits rewriting the exponent in the formula (4.15) in the form

$$\frac{1}{2} \int \frac{\delta^2 S_\Lambda}{\delta A_\mu^a(x) \delta A_\nu^b(y)} q_\mu^a(x) q_\nu^b(y) dx\, dy - g t^{abc} \int \frac{\delta S_\Lambda}{\delta A_\nu^a(x)}$$

$$\times \left[q_\nu - \frac{1}{2} \nabla_\nu \nabla^{-2} \nabla_\rho q_\rho \right]^b [\nabla^{-2} \nabla_\sigma q_\sigma]^c dx. \quad (4.18)$$

The second term in (4.18) can be readily verified to generate only convergent diagrams. This follows from the propagator q_μ behaving at large k like k^{-6} and from the number of derivatives at the interaction vertices being insufficient for compensating this decrease. Therefore insertion of vertices generated by the second term in (4.18) into any diagram makes it finite. Thus, for example, the contribution of this term on the mass shell is just zero.

So, we have shown that up to finite terms the functional (4.7) coincides with the manifestly gauge invariant functional (4.8). Since this functional is invariant under gauge transformations of the fields $\mathcal{A}_\mu$, its local divergent part obviously exhibits the same property also; and, consequently, the counterterms necessary for the removal of ultraviolet divergences coincide with the Yang-Mills Lagrangian

$$\Delta \mathcal{L} = B \frac{1}{8} \mathrm{tr} \mathcal{F}_{\mu\nu} \mathcal{F}_{\mu\nu}. \quad (4.19)$$

We note that such a simple renormalization procedure is characteristic of the Yang-Mills theory with higher derivatives. In the theory, which does not contain higher derivatives, counterterms, as we shall see below, exhibit a more complex structure.

The proof presented above of the invariance of the counterterm part of the one-loop action was of a formal character, since we dealt with divergent integrals. This proof may be rendered quite rigorous by replacing (4.8) with the following similar expression having undergone Pauli-Villars regularization:

$$Z_0(\mathcal{A}) \to Z_0^R(\mathcal{A}) = \det Q_0^{-1/2} \prod_j \det Q_j^{-\frac{c_j}{2}} \det B_0 \det B_j^{c_j}, \tag{4.20}$$

where

$$\begin{aligned}\det B_j &= \det\{\nabla^2 - \mu_j^2\} \\ &= \int \exp\left\{-\frac{i}{2}\int \operatorname{tr}[\bar{b}_j \nabla^2 b_j - \mu_j^2 \bar{b}_j b_j]dx\right\} \prod_x d\bar{b}_j db_j,\end{aligned} \tag{4.21}$$

$$\begin{aligned}\det Q_j^{1/2} = \int \exp\Bigg\{i \int \Bigg[\frac{1}{2}\int \frac{\delta^2 S_\Lambda}{\delta A_\mu^a(x)\delta A_\nu^b(y)} q_{\mu j}^a(x) q_{\nu j}^b(y) \\ - \frac{\mu_j^2}{2} q_{\mu j}^2(x)\delta(x-y)\Bigg] dx\, dy\Bigg\} \prod_x \delta(\nabla_\mu q_{\mu j}) dq_{\mu j}.\end{aligned} \tag{4.22}$$

The fields $q_{\mu j}(\bar{b}_j, b_j)$ are considered to commute when $c_j > 0$ $(c_j < 0)$ and to be anticommuting when $c_j < 0$ $(c_j > 0)$; the coefficients c_j satisfy the Pauli-Villars conditions

$$\Sigma c_j + 1 = 0, \qquad \Sigma c_j \mu_j^2 = 0, \tag{4.23}$$

$$B_0 = B \text{ when } \mu_j = 0, \qquad Q_0 = Q_j \text{ when } \mu_j = 0.$$

The formula (4.21) describes the ordinary Pauli-Villars regularization discussed above in detail, taking advantage of the example of spinor fields. From each closed cycle, along which a vector (q_μ) or a scalar (b) particle propagates, similar cycles are subtracted in which the internal lines possess masses μ_j. Owing to the conditions (4.23), the leading terms in the asymptotics of the integrands cancel out, and the integrals become convergent.

The factors $\det Q_j$ and $\det B_j$ are obviously gauge-invariant. This follows directly from the integral representations (4.21) and (4.22). The corresponding effective actions describe the gauge-invariant interaction of scalar (b) and vector (q_μ) fields with the Yang-Mills field. Therefore

$$\begin{aligned}\det B_j(\mathcal{A}^\omega) &= \det B_j(\mathcal{A}), \\ \det Q_j(\mathcal{A}^\omega) &= \det Q_j(\mathcal{A}).\end{aligned} \tag{4.24}$$

Choosing as the initial expression the regularized expression (4.20), we may apply to it the transformations described above, and as a result, we will obtain an equality between the regularized expressions (up to terms remaining finite in the limit of the removed ultraviolet regularization):

$$Z_0^R = N^{-1} \int \exp\left\{ i \left[\frac{1}{2} \frac{\delta^2 S_\Lambda}{\delta A_\mu^a(x) \delta_\nu^b(y)} q_\mu^a(x) q_\nu^b(y) \right] dx\, dy \right.$$

$$\left. + \frac{1}{2\alpha} \int \{ f(\Box) \partial_\mu q_\mu^a(x) \}^2 dx \right\} \det M(\mathcal{A}) \prod_j \det Q_j^{-\frac{c_j}{2}} \det B_j^{c_j}$$

$$= N^{-1} \det Q_0^{-1/2} \prod \det Q_j^{-\frac{c_j}{2}} \det B_0 \prod_j \det B_j^{c_j}. \quad (4.25)$$

Tending the regularizing masses μ_j toward infinity and equating the terms singular as $\mu_j \to \infty$, we obtain the equality (4.19).

The problem formulated above is, in principle, solved. Since the subtraction procedure is applied recurrently—first to one-loop diagrams, then to two-loop diagrams, etc; we may subtract the counterterms (4.19), upon which the remaining integral will contain no ultraviolet divergences if the regularization parameter Λ is finite. However, for many purposes it is convenient to have a gauge-invariant regularized generating functional containing no ultraviolet divergences before application of the subtraction procedure. An obvious candidate for such a functional is the expression

$$Z_{\Lambda,\mu} = N^{-1} \int \exp\{iS_\Lambda\} \det M \prod_j \det Q_j^{-\frac{c_j}{2}} \det B_j^{c_j} \prod_x d\mathcal{A}_\mu, \quad (4.26)$$

where S_Λ is the gauge-invariant Yang-Mills action containing higher derivatives, and $\det Q_j$ and $\det B_j$ are determined by the formulas (4.21) and (4.22). Indeed, in the limit $\mu_j \to \infty$ the factors $\det Q_j$ and $\det B_j$ generate the necessary one-loop counterterms, and the combinatorics of the R-operation guarantees that upon subtraction of these counterterms all the remaining integrals are finite. In the case of finite μ_j a similar subtraction takes place, only with the difference that the corresponding "counterterms" are not local (Fig. 14).

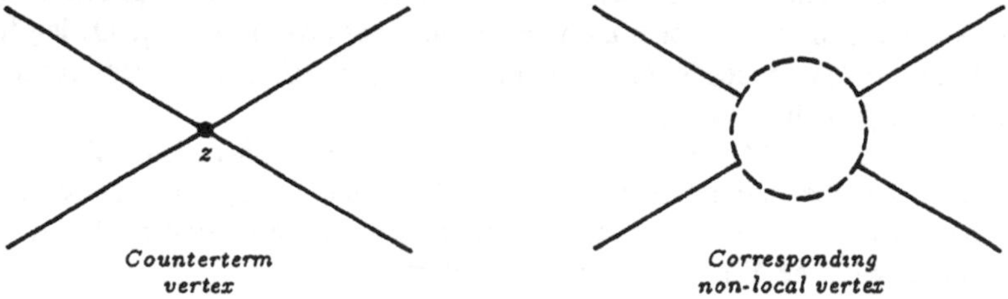

Figure 14. The solid lines indicate the fields A_μ, the dashed lines indicate the fields q_μ and b.

However, besides "counterterm" diagrams, the factors $\det Q_j$ and $\det B_j$ also generate convergent diagrams in which the derivatives act upon the external $\mathcal{A}$-lines but not on the internal q-lines. In the limit $\mu_j \to \infty$ the contribution of these diagrams vanishes. But when the μ_j are finite, these diagrams are present; and upon being integrated over A_μ they could, in principle, generate new ultraviolet divergences. Here we actually encounter overlapping divergences. For this not to happen it is sufficient for the propagators of the fields A_μ to fall off more rapidly than the propagators of the fields q_μ, i.e., the order of the higher derivatives in the regularized action S_Λ must be higher than in the effective action (4.22) determining det Q_i. The structure of one-loop divergences in the Yang-Mills theory with higher derivatives is always determined by the formula (4.19), therefore it is not necessary to consider S_Λ occurring in (4.26) to coincide with S'_Λ determining det Q_j. It is readily shown that by choosing S_Λ and S'_Λ appropriately (for instance, we may consider S_Λ to contain derivatives of the sixth order and S'_Λ to contain derivatives of the fourth order) it is possible to provide for the finiteness of all the diagrams generated by expansion of the regularized functional (4.26) in a perturbation-theory series.

With the aid of the integral representations (4.21) and (4.22) this expression can be represented in the form of the path integral of $\exp\{i \times \text{local action}\}$ over the Yang-Mills fields A_μ and the auxiliary fields $\bar{c}, c, q_{\mu j}, \bar{b}_j, b_j$.

In the case of finite Λ, μ_j, all the diagrams generated by S_Λ converge. At the same time, Z exhibits the same transformation properties that are exhibited formally by the non-regularized functional. Thus, for example, the generalized Ward identities hold for it, as we shall see below.

Removal of the ultraviolet regularization is realized upon introduction of necessary counterterms depending on μ_j, Λ by transition to the limit $\mu_j \to \infty, \Lambda \to \infty$.

4.5 Dimensional Regularization

The divergence index ω depends significantly on the dimensionality of the space-time. For a space of dimension n, the product of independent differentials gives to the diagram index a contribution equal to

$$N(L - m + 1), \tag{5.1}$$

where L is the number of internal lines and m is the number of vertices. Therefore, the diagrams to which divergent integrals correspond in a four-dimensional space may turn out to be convergent in a space of smaller dimensionality. On the other hand, going from the four-dimensional space to an n-dimensional space does not influence the symmetry properties. The gauge transformations are generalized in a natural way to a space of any positive dimension. One can go further and define the Feynman diagrams for spaces of noninteger and even complex dimensions. In this case, of course, one cannot speak about any symmetry of the Lagrangian, since the concept itself loses its sense for a non-integer n. Nevertheless, we can investigate the Green functions in a space of arbitrary dimension. As we shall show below, gauge invariance in terms of the Green functions is equivalent to the existence of relations between these functions, known as the generalized Ward identities. These relations make sense in a space of any dimension; and in the region of n where all the integrals converge, they can be proved rigorously. The Green functions, considered as functions of the space dimensionality n, have pole singularities at $n = 4$. Subtracting these singularities, we can continue the Green functions analytically to the point $n = 4$. The functions obtained in this way will satisfy the genreralized Ward identities.

At the simplest example, consider the integral corresponding to the second-order self-energy diagram

$$I = g^2(\mu^2)^{2-\frac{n}{2}} \int d^n k \frac{1}{(k^2 - m^2 + i0)[(p-k)^2 - m^2 + i0]}. \tag{5.2}$$

In this formula μ is an arbitrary constant of the mass dimension and g is a dimensionless coupling constant. The necessity of introducing the scaling factor μ is due to the fact that for conserving the correct dimensionality of the action in a space of arbitrary dimension, the constants, by which the interaction vertices are multiplied, must be of non-zero dimension.

For integer $n < 4$ this integral converges. Taking advantage of the formula

$$\frac{1}{ab} = \int_0^1 \frac{dx}{[ax + b(1-x)]^2}, \tag{5.3}$$

it can be rewritten in the form

$$I = g^2(\mu^2)^{2-\frac{n}{2}} \int_0^1 dx \int d^n k \frac{1}{[k^2 + p^2 x(1-x) - m^2 + i0]^2}. \tag{5.4}$$

Rotating the contour of integration through 90° and changing the variables, $k_0 \to ik_0$, we arrive at the following integral over the Euclidean space of dimension n:

$$I = ig^2(\mu^2)^{2-\frac{n}{2}} \int_0^1 dx \int d^n k \frac{1}{[k^2 + m^2 - p^2 x(1-x)]^2}. \tag{5.5}$$

The integral over k is readily calculated with the aid of the known formula

$$\int \frac{d^n k}{(k^2 + c)^\alpha} = \frac{\pi^{n/2} c^{\frac{n}{2}-\alpha} \Gamma\left(\alpha - \frac{n}{2}\right)}{\Gamma(\alpha)}, \tag{5.6}$$

where $\Gamma(\alpha)$ is the Euler gamma function. The right-hand side of the equality (5.6) can be analytically continued to complex values of n. We shall treat the formula (5.6) as the definition of the integral on the left-hand side for an arbitrary space dimension. Thus, the integral (5.2) is equal to

$$I = ig^2(\mu^2)^{2-\frac{n}{2}} \pi^{n/2} \Gamma\left(2 - \frac{n}{2}\right) \int_0^1 dx [m^2 - p^2 x(1-x)]^{\frac{n}{2}-2} \tag{5.7}$$

At $n = 4$ the Γ-function has a pole and I tends to infinity. This corresponds to the divergence of the original integral over the four-dimensional space. Expanding I

in the Laurent series around the point $n/2$, we have

$$I = \frac{i\pi^2 g^2}{2-\frac{n}{2}} + i\pi^2 g^2 \varphi(1) - i\pi^2 g^2 \int_0^1 dx \ \ln \frac{[m^2 - p^2(1-x)x]}{\mu^2} + 0\left(\frac{n}{2} - 2\right). \quad (5.8)$$

Here $\psi(\alpha) = \frac{d \ln \Gamma(\alpha)}{d\alpha}$.

In accordance with the general rules formulated above for the removal of divergences, the renormalized self-energy can be defined with the aid of a subtraction procedure. Performing subtraction at the point $\lambda^2 = p^2$ we obtain the final result in the form

$$I_R = i\pi^2 g^2 \int_0^1 dx \ \ln \frac{m^2 - \lambda^2 x(1-x)}{m^2 - p^2 x(1-x)}. \quad (5.9)$$

Other divergent diagrams can be redefined in a similar manner. It is often convenient, however, to apply a special renormalization procedure, which has been termed the minimal-subtraction scheme. The subtraction procedure, in this case, reduces to discarding the pole term in the expression (5.8). In other words, the renormalized self-energy in the minimal-subtraction scheme is given by the expression

$$I_{R,\min} = i\pi^2 g^2 \psi(1) - i\pi^2 g^2 \int_0^1 dx \ \ln \frac{[m^2 - p^2(1-x)x]}{\mu^2}. \quad (5.10)$$

A similar recipe is adopted for all the remaining divergent diagrams, too. Naturally, from the point of view of any individual diagram this recipe reduces to the choice of a particular subtraction point. Thus, for example, this point for the diagram (5.2) is determined from the equation

$$\psi(1) = \int_0^1 dx \ \ln \frac{[m^2 - \lambda^2 x(1-x)]}{\mu^2}. \quad (5.11)$$

If, however, one deals with the entire set of divergent diagrams, no universal choice of the normalization point can any longer be found to correspond to the minimal-subtraction procedure. Various subtraction points are required for various diagrams, and even for diagrams of identical structure (for instance, in the case of the energy parts proper) the subtraction points alter depending on the order.

The convenience of the minimal-subtraction procedure is due to the counterterms exhibiting, in this case, an especially simple structure; thus, for instance, they are totally independent of mass. This essentially simplifies many calculations, in particular, solution of the renormalization group equations to be considered in the last chapter. Besides this, if there exist any symmetry relations between regularized Green functions (such as, for example, the generalized Ward identities), they are automatically conserved, also, for the renormalized quantities. This follows from the symmetry relations having to be satisfied for each term of the Laurent expansion of the corresponding Green functions in the vicinity of the point $n = 4$. Therefore discarding the pole terms cannot violate these relations.

For calculations of arbitrary Feynman diagrams it is necessary to formulate also rules for treating tensor entities in an n-dimensional space. By definition,

$$g_{\mu\nu}p_\nu = p_\mu; \quad p_\mu p_\mu = p^2; \quad g_{\mu\nu}g_{\nu\alpha} = \delta_{\mu\alpha}; \quad g_{\mu\nu}g_{\mu\nu} = n. \tag{5.12}$$

In a similar manner, for theories including fermions, objects are introduced which possess the algebraic properties of the γ-matrices:

$$\gamma_\mu\gamma_\nu + \gamma_\nu\gamma_\mu = 2g_{\mu\nu}I, \tag{5.13}$$

where I is the identity matrix, and

$$\mathrm{tr}\{\gamma_\mu\gamma_\nu\} = 2^{\frac{n}{2}}g_{\mu\nu}, \tag{5.14}$$

$$\gamma_\mu\hat{p}\gamma_\mu = 2\left(1 - \frac{n}{2}\right)\hat{p}, \quad \gamma_\mu\hat{p}\hat{q}\gamma_\mu = 4pq + (n-4)\hat{p}\hat{q}. \tag{5.15}$$

Note, however, that the usual definition of the matrix γ_5,

$$\gamma_5 = \frac{-i}{4!}\varepsilon_{\alpha\beta\mu\nu}\gamma_\alpha\gamma_\beta\gamma_\mu\gamma_\nu \tag{5.16}$$

is not applicable in a space of arbitrary dimension since the completely antisymmetric tensor $\varepsilon_{\alpha\beta\mu\nu}$ is defined only in the four-dimensional space. Because of this, the theories to which the matrix γ_5 pertains need special consideration, and in general the dimensional regularization is not applicable to them.

The recipe for the dimensional regularization of an arbitrary Feynman diagram consists in the following. First of all, the tensor structures are singled out of the γ-matrices belonging to the external fermion lines. With the aid of the formal algebra of γ-matrices in an n-dimensional space, described above, one further calculates the traces of products of the γ-matrices corresponding to the internal fermion lines. As a result, there arise integrals of the form

$$F = \prod_{j=1}^{I} d^n k_j (k_{l_1})_\lambda (k_{l_2})_\mu \ldots (k_{l_a})_\nu \prod_{i=1}^{L} (q_i^2 - m_i^2 + i0), \tag{5.17}$$

where L is the number of internal lines in the diagram, I is the number of independent cycles, and the momenta q_i represent algebraic sums of the integration variables k_i and the external momenta p_i. In these integrals it is necessary to use some parametrization allowing us to perform integration over the angular variables explicitly. To this end it is possible to use either the Feynman parametrization with subsequent transition to the Euclidean region, as done in the example (5.2), or the so-called α-representation

$$(p^2 - m^2 + i0)^{-1} = (i)^{-1}\int_0^\infty d\alpha \exp\{i\alpha(p^2 - m^2 + i0\}. \tag{5.18}$$

After passing to the α-representation, the integrals over k become Gaussian and are calculated by formulas of the type

$$\int \frac{d^n k}{(2\pi)^n} \exp(-xk^2 + 2ka) = \left(\frac{\pi}{x}\right)^{\frac{n}{2}} (2\pi)^{-n} \exp\left\{\frac{a^2}{x}\right\}. \tag{5.19}$$

For noninteger n, the formula (5.19) is considered to be the definition of the integral over the n-dimensional space. The integral defining the function F converges in a finite region of the complex variable n. At $n = 4$ this function has poles. (In practice, these poles appear, for example, as singularities of the Euler Γ-functions that result from integration over the parameters α.)

The funtions F can be represented in the form of a product of tensors constructed of the external momenta p_i and scalar functions F_j. If the corresponding diagram has no divergent subgraphs, then the Laurent expansion of the functions F_j around the point $n = 4$ has the form

$$F_i(p) = \frac{A(p_i^2, p_i p_j)}{(n-4)} + B(p_i^2, p_i p_j) + 0(n-4), \tag{5.20}$$

where $A(p_i^2, p_j p_j)$ is a polynomial of an order equal to the diagram index.

We note that, as it follows for the formula (5.6),

$$\int \frac{d^4 k}{(k^2)^\alpha} = 0 \quad \text{for } \alpha < 2 \tag{5.21}$$

therefore in the method of dimensional regularization there exist no so-called quadratic divergences, i.e., no counterterms that, in the renormalization procedure involving the ultraviolet cut-off Λ, are proportional to Λ^2.

The next step, the renormalization procedure, depends on which subtraction procedure we wish to apply.

In the minimal-subtraction scheme one must drop the pole terms in the expression for F_j and consider the remaining finite expression to be the definition of the renormalized function F_j.

The alternative method of renormalizaton consists in subtracting from $F_j(p)$ several of the first terms of the expansion in a Taylor series in the momenta p at some point λ, for instance, on the mass shell (if possible).

For diagrams containing divergent subgraphs, the subtractions are performed successively. Counterterms are introduced, which remove the divergence from the subgraphs, and then a subtraction is performed for the diagram as a whole. An important property of the dimensional regularization is the possibility of shifting the integration variables within regularized integrals. It is just this property, together with the tensor algebra (5.12), which allows one to prove the generalized Ward identities within the framework of dimensional regularization.

In conclusion we shall illustrate the dimensional-regularization method by a simple example—the calculation of second-order corrections to the Green function of the Yang-Mills field. This correction is described by the diagrams in Fig. 9.

To the diagram (a) in the diagonal α-gauge, there corresponds the regularized integral

$$\Pi^{ab}_{\mu\nu}(p)_a = -\frac{g_1^2}{2}\int\frac{d^n k}{(2\pi)^n}\varepsilon^{aa_1a_2}\varepsilon^{bb_1b_2}$$
$$\times[(p+k)_{\mu_2}g_{\mu\mu_1}+(p-2k)_\mu g_{\mu_1\mu_2}+(k-2p)_{\mu_1}g_{\mu\mu_2}]$$
$$\times[(k+p)_{\nu_2}g_{\nu\nu_1}+(k-2p)_{\nu_1}g_{\nu\nu_2}+(p-2k)_\nu g_{\nu_1\nu_2}]$$
$$\times\frac{(-i\delta^{a_1b_1})}{k^2+i0}g_{\mu_1\nu_1}\frac{(-i\delta^{a_2b_2})}{(p-k)^2+i0}g_{\mu_2\nu_2}$$
$$= g_1^2\delta^{ab}\int\frac{d^n k}{(2\pi)^n}\{g_{\mu\nu}[(k+p)^2+(k-2p)^2]+(n-6)p_\mu p_\nu$$
$$+(4n-6)k_\mu k_\nu+(3-2n)(p_\nu k_\mu+p_\mu k_\nu)\}\times\{(k^2+i0)[(p-k)^2+i0]\}^{-1}, \quad (5.22)$$

where, in order to conserve the correct dimensionality of $\Pi_{\mu\nu}$ a dimensional coupling constant $g_1^2 = g^2\mu^{4-n}$ has been introduced.

Using the formula

$$\frac{1}{k^2(p-k)^2} = \int_0^1 dz\frac{1}{[k^2(1-z)+(p-k)^2 z]^2} \quad (5.23)$$

and passing to new variables

$$k \to k + pz, \quad (5.24)$$

we write this integral in the form

$$\Pi^{ab}_{\mu\nu}(p)_a = g_1^2\delta^{ab}\int_0^1 dz\frac{d^n k}{(2\pi)^n}[g_{\mu\nu}(5-2z+2z^2)p^2+2g_{\mu\nu}k^2$$
$$+(4n-6)k_\mu k_\nu-(4n-6)z(1-z)p_\mu p_\nu$$
$$+(n-6)p_\mu p_\nu][k^2+p^2 z(1-z)+i0]^{-2}. \quad (5.25)$$

In this formula the terms odd in k are dropped, since their contribution vanishes for symmetry reasons. Passing to the Euclidean metric, it is possible to integrate over k by the formula

$$\int\frac{(k^2)^m d^n d}{[k^2+p^2 z(1-z)]^l} = \frac{i\pi^{n/2}}{\Gamma\left(\frac{n}{2}\right)}\int_0^\infty dk^2\frac{(-k^2)^m(k^2)^{\frac{n}{2}-1}}{[-k^2+p^2 z(1-z)]^l}$$
$$= \frac{i\pi^{\frac{n}{2}}}{\Gamma\left(\frac{n}{2}\right)}(-1)^{m+l}[-p^2 z(1-z)]^{m+\frac{n}{2}-1}$$
$$\times\frac{\Gamma\left(m+\frac{n}{2}\right)\Gamma\left(l-\frac{n}{2}-m\right)}{\Gamma(l)}. \quad (5.26)$$

For a noninteger or complex n this formula represents the definition of the integral in the left-hand side of (5.26). Integrating, we obtain

$$\Pi^{ab}_{\mu\nu}(p)_a = \frac{ig_1^2\delta^{ab}}{(4\pi)^{\frac{n}{2}}}\int_0^1 dz\{[g_{\mu\nu}(5-2z+2z^2)p^2$$
$$-(4n-6)p_\mu p_\nu z(1-z)+(n-6)p_\mu p_\nu]\times[-p^2z(1-z)]^{\frac{n}{2}-2}\Gamma\left(2-\frac{n}{2}\right)$$
$$-3(n-1)g_{\mu\nu}[-p^2z(1-z)]^{\frac{n}{2}-1}\Gamma\left(1-\frac{n}{2}\right)\} \tag{5.27}$$

Integration over z is performed using the formula

$$\int_0^1 dz z^{m-n-1}(1-z)^{m-k-1} = \frac{\Gamma(m-n)\Gamma(m-k)}{\Gamma(2m-n-k)}. \tag{5.28}$$

As a result we obtain

$$\Pi^{ab}_{\mu\nu}(p)_a = \frac{ig_1^2}{(4\pi)^{n/2}}\delta^{ab}\left\{g_{\mu\nu}p^2\left[5\frac{\Gamma^2\left(\frac{n}{2}-1\right)}{\Gamma(n-2)}\right.\right.$$
$$-2\frac{\Gamma\left(\frac{n}{2}\right)\Gamma\left(\frac{n}{2}-1\right)}{\Gamma(n-1)}+2\frac{\Gamma\left(\frac{n}{2}+1\right)\Gamma\left(\frac{n}{2}-1\right)}{\Gamma(n)}$$
$$\left.+\frac{6(n-1)}{2-n}\frac{\Gamma^2\left(\frac{n}{2}\right)}{\Gamma(n)}\right]-p_\mu p_\nu\left[(4n-6)\frac{\Gamma^2\left(\frac{n}{2}\right)}{\Gamma(n)}\right.$$
$$\left.\left.-(n-6)\frac{\Gamma^2\left(\frac{n}{2}-1\right)}{\Gamma(n-2)}\right]\right\}\Gamma\left(2-\frac{n}{2}\right)(-p^2)^{\frac{n}{2}-2}, \tag{5.29}$$

In deriving this formula we have used the relation

$$\Gamma(1-\omega) = \frac{1}{(1-\omega)}\Gamma(2-\omega). \tag{5.30}$$

As is seen, in the limit $n \to 4$,

$$\Pi_{\mu\nu}(p)_a \to \infty,$$

since the function $\Gamma(2-n/2)$ has a pole at this point. Expanding $\Pi_{\mu\nu}(p)_a$ around $n=4$ in the Laurent series and taking into account that

$$\left(\frac{\mu^2}{-p^2}\right)^\varepsilon = 1+\varepsilon\ln\left(\frac{\mu^2}{-p^2}\right)+0(\varepsilon^2), \tag{5.31}$$

we obtain the final expression for $\Pi_{\mu\nu}(p)_a$:

$$\Pi^{ab}_{\mu\nu}(p)_a = \frac{ig^2\delta^{ab}}{16\pi^2}\left\{(g_{\mu\nu}p^2 - p_\mu p_\nu)\left(\frac{19}{6}\varepsilon^{-1} + c_a\right) - \frac{1}{2}p_\mu p_\nu(d+\varepsilon^{-1})\right.$$
$$\left. +(g_{\mu\nu}p^2 - p_\mu p_\nu)\frac{19}{6}\ln\frac{\mu^2}{-p^2} - \frac{1}{2}p_\mu p_\nu \ln\frac{\mu^2}{-p^2}\right\}, \tag{5.32}$$
$$\varepsilon = \frac{4-n}{2},$$

where c_a and d are finite constants.

The integral

$$\Pi^{ab}_{\mu\nu}(p)_b = -2g_1^2 \int \frac{d^n k}{(2\pi)^n}\frac{k_\mu(k-p)_\nu}{k^2(p-k)^2}\delta^{ab}. \tag{5.33}$$

corresponds to the diagram (b). Calculations completely analogous to the preceding ones give

$$\Pi^{ab}_{\mu\nu}(p)_b = \frac{ig^2}{16\pi^2}\delta^{ab}\left\{(g_{\mu\nu}p^2 - p_\mu p_\nu)\left(\frac{1}{6}\varepsilon^{-1} + c_b\right) i\right.$$
$$\left. +\frac{1}{2}p_\mu p_\nu(\varepsilon^{-1} + d) + (g_{\mu\nu}p^2 - p_\mu p_\nu)\frac{1}{6}\ln\frac{\mu^2}{-p^2} + \frac{1}{2}p_\mu p_\nu \ln\frac{\mu^2}{-p^2}\right\}. \tag{5.34}$$

And, finally, the diagram (c) gives a zero contribution. The contribution of this diagram is proportional to the integral

$$I = \int \frac{d^4 k}{k^2}. \tag{5.35}$$

In the method of dimensional regularization the formula

$$I = \int \frac{d^n k (k^2)^{\alpha-1}}{(2\pi)^n} = 0; \tag{5.36}$$
$$\alpha = 0, 1, \ldots, m.$$

is valid. Thus, the total second-order correction to the Green function of the Yang-Mills field has the form

$$\Pi^{ab}_{\mu\nu}(p) = \Pi^{ab}_{\mu\nu}(p)_a + \Pi^{ab}_{\mu\nu}(p)_b = \frac{ig^2\delta^{ab}}{16\pi^2}(g_{\mu\nu}p^2 - p_\mu p_\nu)\left\{\frac{10}{3}\varepsilon^{-1} + c + \frac{10}{3}\ln\frac{\mu^2}{-p^2}\right\}. \tag{5.37}$$

The divergence as $\varepsilon \to 0$ is removed, as usual, by a subtraction procedure. The corresponding counterterm is equal to

$$(z_2 - 1) = \frac{5g^2}{24\pi^2}\varepsilon^{-1} + b, \tag{5.38}$$

where b is an arbitrary constant. In the minimal-subtraction scheme $b = 0$. This result is in agreement with the formula (1.33) from Section 4.1. As is seen, the expression (5.37) is automatically transverse, and for the removal of divergences there is no need of gauge-noninvariant counterterms, such as the counterterm for the mass renormalization of the Yang-Mills field.

4.6 Gauge Fields in Lattice Space-Time

In this section, we shall discuss one more method of gauge-invariant regularization of the Yang-Mills theory by substitution of a discrete lattice for the continuous space-time. Such regularization seems to be most natural from the point of view of the path-integral formalism since the path intgral is introduced as the limit of finite-dimensional approximations. However, for the purposes of the Feynman diagram technique, which is the main point at issue in this book, lattice regularization is inconvenient owing to the absence of manifestly relativistic invariance and to computational difficulties. At the same time this regularization is applied within certain approaches, which are not related to expansion in the coupling constant, and, especially, in numerical computer calculations. A detailed discussion of such methods is outside the scope of our book, and we refer the interested reader to the relevant reviews and monographs. Here we shall restrict ourselves to formulating the general scheme and to briefly touching upon possible applications unrelated to perturbation theory.

There exist two approaches to the lattice regularization of gauge theories. In one of them, the theory is, from the very beginning, considered in the Euclidean space R^4, and the lattice is introduced along all four directions. Transition to pseudo-Euclidean space is accomplished by analytical continuation. In the second approach, the theory is considered in the pseudo-Euclidean space, and only the spatial variables are discretized, while time remains continuous.

Within the first approach, field theory essentially reduces to four-dimensional statistical mechanics, which permits application of methods developed in this field of science.

In the second approach, one deals with a quantum-mechanical system involving a large number of degrees of freedom. Both approaches exhibit advantages and disadvantages. The most commonly applied one is the Euclidean formulation, which we shall now proceed to discuss.

The continuous Euclidean space R^4 is replaced with a discrete lattice Z^4. For definiteness, we shall consider a regular cubic lattice with a lattice constant a. We shall denote the vertices of the lattice by $n = (n_1, n_2, n_3, n_4)$, where the n_i are integer numbers. The Fourier transform for a function defined on the lattice is given by the usual formula

$$\tilde{f}(p) = a^4 \sum_n e^{i(pn)a} f(n). \tag{6.1}$$

Since $\tilde{f}(p)$ is a periodic function of p_μ with a period $\frac{2\pi}{a}$, it is sufficient to consider values of p_μ lying within the interval $(\frac{-\pi}{a}, \frac{\pi}{a})$. In this way, the ultraviolet cutoff is introduced into the theory: $|p_\mu| \leq \frac{\pi}{a}, \mu = 1,2,3,4$. The fields of matter are defined naturally at the lattice points: $\varphi(x) \to \varphi(n)$, while vector fields, which are also characterized by direction, are defined on the links (n, μ). Here the point n indicates the beginning of the link, and μ is a unit vector pointing in the direction μ.

Replacing the derivatives in the Lagrangian function with finite differences,

$$\partial_\mu f(x) \to \Delta_\mu f(x) = \frac{1}{a}(f(n+\mu) - f(n)), \tag{6.2}$$

one can readily write the lattice analog of the Yang-Mills action. However, niave discretization of the Yang-Mills Lagrangian would lead to the loss of its most important property, its gauge invariance. To avoid this, one can take advantage of the following device. Instead of the Yang-Mills field $\mathcal{A}_\mu(n)$, assuming its value in the Lie algebra of the compact group Ω, we shall consider the function $U_{n,\mu}$, which assumes values within the group Ω. It is convenient to associate $U_{n,\mu}$ with the link starting at the point n and terminating at the point $n + \mu$. The inverse element is associated with the link pointing in the opposite direction,

$$U_{n,\mu}^{-1} = U_{n+\mu,-\mu}. \tag{6.3}$$

Under the transformations from the gauge group $U_{n,\mu}$ transforms as allows:

$$U_{n,\mu} \to \Omega_n U_{n,\mu} \Omega_{n+\mu}^{-1}. \tag{6.4}$$

where Ω_n is a function defined on the lattice and possessing values in the group Ω.

From a geometrical point of view the element $U_{n,\mu}$ realizes parallel translation along the link μ, while Ω_n defines local rotation of the basis at the point n. In the continuum limit $a \to 0$ the set Ω_n transforms into the element $\Omega(x)$ of the gauge group, while the gauge field $\mathcal{A}_\mu(x)$ with values in the Lie algebra of the group Ω appears under the exponential parametrization of $U_{n,\mu}$:

$$U_{n,\mu} = \exp\{-a\mathcal{A}_\mu(n)\} \tag{6.5}$$

One can readily verify that in the limit $a \to 0$ the transformations (6.4), being written in terms of the fields $\mathcal{A}_\mu$, transform into conventional gauge transformations. Indeed,

$$1 - a\mathcal{A}_\mu(n) + \ldots \to \Omega_n(1 - a\mathcal{A}_\mu(n))(\Omega_n + a\Delta_\mu\Omega_n)^{-1} + \ldots \tag{6.6}$$

from which, upon equating the first-order terms in a, we obtain

$$\mathcal{A}_\mu \to \Omega\mathcal{A}_\mu\Omega^{-1} + \partial_\mu\Omega\Omega^{-1}. \tag{6.7}$$

The simplest action invariant with rspect to the transformations (6.4) has the form

$$S = \sum_p \frac{1}{g^2}\operatorname{tr}[U_p + U_p^+ - 2], \tag{6.8}$$

where

$$U_p = U(p_{n,\mu\nu}) = U_{n,\mu}U_{n+\mu,\nu}U_{n+\mu+\nu,-\mu}U_{n+\mu,-\nu}, \tag{6.9}$$

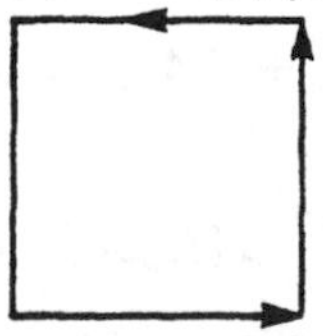

Figure 15. The plaquette $p_{n,\mu\nu}$.

and $p_{n,\mu\nu}$ is a plaquette formed by the links $(n,\mu), (n+\mu,\nu)$, $(n+\mu+\nu,-\nu)$, $(n+\nu,-\nu)$ (Fig. 15).

The quantity $U(p_{n,\mu\nu})$ indicates parallel translation along the plaquette in the given direction. Summation is performed over all plaquettes, while U and U^+ correspond to plaquettes with opposite orientation.

It may be readily verified that in the continuum limit (6.8) transforms into the Yang-Mills action. Applying the Baker-Hausdorf formula, we obtain

$$U(p) = 1 + a^2\mathcal{F}_{n,\mu\nu} + \frac{1}{2!}a^4\mathcal{F}^2_{n,\mu\nu} + \ldots, \tag{6.10}$$

where

$$\mathcal{F}_{n,\mu\nu} = \Delta_\mu \mathcal{A}_{n\nu} - \Delta_\nu \mathcal{A}_{n\mu} + [\mathcal{A}_{n\mu}, \mathcal{A}_{n\nu}] + O(a). \tag{6.11}$$

The contribution of the term proportional to a^2 cancels out when the summation $U(p)+U^+(p)$ is performed, and the term of the order of magnitude of a^4 transforms in the limit $a \to 0$ into the continuous Yang-Mills action.

The vacuum expectation values corresponding to the Green functions of the continuous theory are given by the path integral over the invariant measure on the group Ω:

$$\begin{aligned} \langle f(U_{n,\mu})\rangle &= \frac{1}{N}\int f(U_{n,\mu})e^{-S(U)}\prod_{n,\mu} d\mu(U_{n,\mu}), \\ N &= \int e^{-S(U)}\prod_{n,\mu} d\mu(U_{n,\mu}). \end{aligned} \tag{6.12}$$

Note that in the lattice theory the choice of the quantities U as the principal variables is determined by reasons of convenience. If necessary, transition can be performed to the variables $\mathcal{A}_{n,\mu}$ by the substitution of variables (6.5). This is precisely what has to be done if one wishes to compute integrals such as (6.12) applying perturbation theory in the coupling constant g. Redefining in the conventional manner the variables, $\mathcal{A}_{n,\mu} \to g\mathcal{A}_{n,\mu}$, we can expand $U_{n,\mu}$ in a series of perturbation theory:

$$U_{n,\mu} \approx 1 - ga\mathcal{A}_{n,\mu} + \ldots \tag{6.13}$$

The corresponding expansion for the action has the form

$$S = a^4 \sum_{n,\mu\nu} \frac{1}{8}\mathrm{tr}\,(\Delta_\nu A_{n\mu} - \Delta_\mu A_{n\nu} + g[A_{n\mu}, A_{n\nu}])^2 - \frac{ag}{4}\mathrm{tr} A_{n\mu}[\Delta_\nu A_{n\mu}, \Delta_\mu A_{n\nu}]$$

$$-\frac{ag^2}{4}\mathrm{tr}\,[\Delta_\mu A_{n\nu}, A_{n\mu}][A_{n\mu}, A_{n\nu}] + \ldots \tag{6.14}$$

By fixing, as usual, the gauge it is possible to construct the perturbation theory starting from the action (6.14).

However, in this action, besides the vertices that in the limit $a \to 0$ transform into standard Yang-Mills vertices, there are new vertices involving a large number of fields and additonal derivatives. Although these vertices are proportional to positive powers of the lattice constant a, and in classical theory they are insignificant in the limit $a \to 0$, they cannot be neglected in the quantum case. Since these vertices contain additional derivatives or higher powers of the field, the corresponding diagrams diverge inthe limit $a \to 0$ as a^{-k} and, as a result, yield a non-zero contribution. As the order of perturbation theory increases, there appear more and more new vertices, which renders the lattice regularization quite unsuitable from the point of view of practical calculations. Besides this, owing to the absence of explicit relativistic invariance, relativistically noninvariant counterterms may be required for the removal of ultraviolet divergences.

Therefore, in practice, this regularization is rarely applied in perturbation theory, and the main scope of its applications are nonperturbative methods. To conclude this section, we shall briefly touch upon these methods.

For computation of the correlation functions (6.12) one may take advantage of the strong-coupling expansion, i.e., perturbation theory in $1/g^2$. As an example, we shall consider the calculation of the vacuum expectation value of the so-called Wilson loop, which in the continuous case is determined by the integral

$$\langle W(\Gamma)\rangle = \int \frac{1}{N}\mathrm{tr} P \exp \int_\Gamma A_\mu dx_\mu \cdot \exp\{-S\} \prod_{x,\mu} dA_\mu(x). \tag{6.15}$$

Here Γ is a closed contour, and P indicates ordering along the contour. In lattice theory the following regularized integral corresponds to this vacuum expectation value:

$$\langle W(\Gamma)\rangle = \frac{1}{W_0} \int \mathrm{tr}\,(U_{n_1,\mu_1} U_{n_1+\mu_1,\mu_2} \ldots U_{n_1-\mu_n,\mu_n})$$

$$\times \exp\left\{-\frac{1}{8g^2}\sum_p \mathrm{tr}\,(U_p^+ + U_p - 2)\right\} dU,$$

$$W_0 = \int dU \exp\left\{-\frac{1}{8g^2}\sum_p \mathrm{tr}\,(U_p^+ + U_p - 2)\right\}. \tag{6.16}$$

The numerator in this formula contains the ordered product of group factors U_{n_i,μ_i} corresponding to the lattice links forming the contour Γ (Fig. 16).

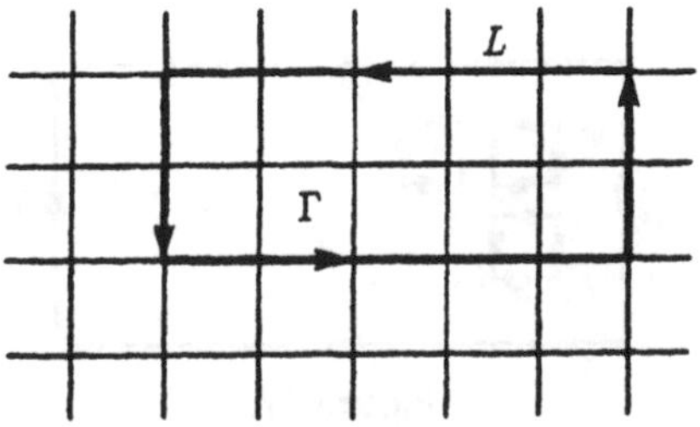

Figure 16. A Wilson loop on the lattice.

We shall calculate the integral (6.16) by expanding the exponential in powers of $1/g^2$. As one can see, the problem reduces to the computation of integrals such as

$$\int U_{i_1 j_1} \dots U_{i_n j_n} U^+_{k_1 l_1} \dots U^+_{k_m l_m} d\mu(U), \tag{6.17}$$

where i_n, j_n, k_m, l_m are group indices. These integrals are calculated with the aid of standard formulas for integration on a group. Thus, for the $U(N)$ group we have

$$\begin{gathered} \int d\mu(U)U = \int d\mu(U)U^+ = \int d\mu(U)U^+U^+ = \int d\mu(U)UU = 0 \\ \int d\mu(U)U_{ij}U^+_{kl} = \frac{1}{N}\delta_{il}\delta_{jk}. \end{gathered} \tag{6.18}$$

The integral (6.17) differs from zero only when for each U_{ij} present in the product there can be found a corresponding factor U^+_{ji} possessing the same indices. Geometrically this means that a contribution differing from zero is given by those products U in which each lattice link participates twice, once being taken in the direct direction and the second time in the opposite direction. Returning to the integral (6.16) we see that the leading order in $1/g^2$ corresponds to the term in which the plaquettes U_p appearing in the expansion of the exponential fill up the surface of the Wilson loop in a minimal manner (Fig. 17). Therefore, in the leading order in $1/g^2$,

$$\langle W(\Gamma)\rangle \sim \left(\frac{1}{Ng^2}\right)^{LT} = e^{-LT\ln(Ng^2)}. \tag{6.19}$$

In the case of an arbitrary contour,

$$\langle W(\Gamma)\rangle \sim e^{-S_{\min}(\Gamma)\ln(Ng^2)}, \tag{6.20}$$

where $S_{\min}$ is the area of the minimum surface stretched over the contour Γ.

It can be demonstrated that the vacuum expectation value $\langle W(\Gamma)\rangle$ describes the interaction energy between two separated static sources, namely,

$$\langle W\rangle \underset{T\to\infty}{\sim} e^{-TE(L)}, \tag{6.21}$$

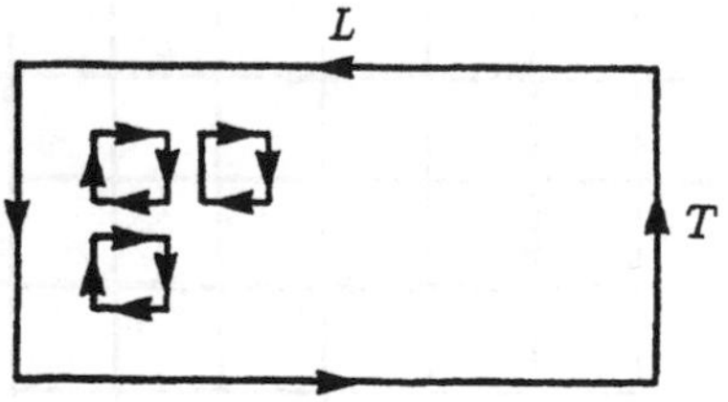

Figure 17.

where $E(L)$ is the interaction energy. As can be seen from formula (6.19), this energy in the Yang-Mills theory has the form

$$E \sim L \ln(Ng^2) \tag{6.22}$$

i.e., a linear interaction potential arises between two static sources. As $g \to \infty$, the charges are bound by an elastic string under a tension equal (in terms of a^{-2}) to

$$\sigma = \ln(3g^2). \qquad (\text{as } N = 3). \tag{6.23}$$

As a result, the distance between them cannot become macroscopic, and they form a permanently bound system. This phenomenon has been termed the confinement of charges.

Regretfully, the behavior of a system as $g \to \infty$ is of little interest from the point of view of plausible models. The point is that ultimately we are interested in a theory in continuous space-time, i.e., in the limit $a \to 0$. In this case, in the limit $a \to 0$, Lorentz invariance must be restored, and an appropriate choice of the constant g. We shall consider some Green function $G(x, y) = \langle A(x)B(y)\rangle$. At large (but finite!) Euclidean $x - y$ the asymptotic behaviour of $G(x - y)$ can naturally be expected to be of the type

$$G(x, y) \sim e^{-\frac{|n-m|}{\xi(g)}} = e^{-\frac{|x-y|}{\xi a}}. \tag{6.24}$$

To obtain a nontrivial result under transition to continuous theory, the correlation length $\xi(g)$ must tend toward infinity as $a \to 0$. It is well known that the correlation length tends to infinity at the points of phase transition. Therefore, to obtain a physically meaningful theory, one must perform transition to the limit in the vicinity of the phase transition point g^*, i.e.,

$$\lim g(a) \underset{a \to 0}{=} g^*. \tag{6.25}$$

In the Yang-Mills theory the critical value of the charge equals zero (see Section 2 of Chapter 5, concerning this issue). Therefore, the limit transition to continuous space-time must take place in the vicinity of the point $g = 0$ where the strong-coupling expansion is certainly not applicable.

At the same time, the fact that the critical value of the coupling constant in the Yang-Mills theory turns zero makes possible important predictions concerning the scaling properties of this theory.

Since the Yang-Mills Lagrangian contains no dimensional parameters, all dimensional quantities, such as, for example, mass, must be expressed in terms of the lattice constant a:

$$m = \frac{1}{a} f(g). \tag{6.26}$$

If the limit transition $a \to 0$ to continuous space-time takes place at the same time as the coupling constant tends to its critical value $g^*(g^* = 0$ for the Yang-Mills field), then m tends toward a finite value equal to the physical mass. Thus, when simultaneously $a \to 0$ and $g \to 0$, the dependence on a in the equation (6.26) vanishes, which is expressed by the following differential equation:

$$\frac{dm}{da} = -\frac{1}{a^2} f(g) + \frac{1}{a}\frac{df}{dg} \cdot \frac{dg}{da} = 0. \tag{6.27}$$

The quantity

$$-\frac{1}{a}\frac{dg}{da} = \beta(g) \tag{6.28}$$

is called the Gell-Mann-Low function and in the vicinity of the point $g = 0$ can be calculated applying perturbation theory (see Section 2 of Chapter 5):

$$\beta(g) = -\beta_0 g^3 - \beta_1 g^5 + 0(g^7), \tag{6.29}$$

where for the Yang-Mills theory in vacuum, corresponding to the $SU(N)$ gauge group,

$$\begin{aligned} \beta_0 &= \frac{11}{3} N \cdot \frac{1}{16\pi^2}, \\ \beta_1 &= \frac{17}{3} N^2 \frac{1}{(16\pi^2)^2}. \end{aligned} \tag{6.30}$$

The solution of the equation (6.27) has the form

$$f(g) = c e^{-\frac{1}{2\beta_0 g^2}} \cdot (\beta_0 g^2)^{-\frac{\beta_1}{2\beta_0^2}}, \tag{6.31}$$

where c is the constant of integration. Hence it follows that in the limit $a \to 0$ and $g \to 0$ the dependence of all the masses and of the other dimensional parameters upon the coupling constant exhibits a universal nature determined by the formula (6.31). Various dimensional quantities differ in the constants c that cannot be calculated by perturbation theory.

There also exist other methods of computation in the lattice Yang-Mills theory such as, for example, the mean-field method or expansion in the parameter $1/N$ for the $SU(N)$ gauge group. Application of lattice regularization, however, became the most popular method for computer Monte-Carlo calculations. Here we shall not discuss this field. It will be sufficient to merely note that, in spite of certain successes, in this field there also exists a series of unsolved problems. One of the difficulties that have not yet been overcome consists in the inclusion into the lattice scheme of fermion fields of matter, which is necessary for plausible physical applications.

4.7 Generalized Ward Identities

We shall now return to the investigation of the Yang-Mills field within the framework of perturbation theory.

A necessary stage in the construction of the quantum scattering matrix, as already pointed out, consists in establishing the renormalization procedure. The renormalization procedure is usually formulated in terms of the Green functions. Unlike the S-matrix, the Green functions are not gauge-invariant objects, and their values depend on the specific gauge condition chosen. The relativity principle is equivalent to the existence of relations between the Green functions, which by analogy with electrodynamics, we shall call the generalized Ward identities. These relations provide for the physical equivalence of various gauges and play a key role in the proof of the gauge invariance and the unitarity of the renormalized S-matrix. From them it follows, in particular, that the counterterms needed for the removal of the intermediate regularization form a gauge-invariant structure.

We shall start with the derivation of the generalized Ward identities for regularized nonrenormalized Green function. In all further reasoning only the gauge invariance of the regularized action will be used. We shall not, therefore, write out its explicit expression, having in mind that we can always use for this purpose, for example, the formula (4.26).

As the initial representation of the generating functional for the Green functions, we shall choose

$$Z = N^{-1} \int \exp\left\{ i \left[S_\Lambda - \frac{1}{2} \operatorname{tr} \int \left[\frac{1}{2\alpha} (f(\Box) W(x))^2 + \mathcal{Y}_\mu \mathcal{A}_\mu \right] dx \right] \right\}$$
$$\times \Delta(\mathcal{A}) \prod_x \delta(\partial_\mu \mathcal{A}_\mu - W) d\mathcal{A} dW. \tag{7.1}$$

Here S_Λ is the gauge-invariant action functional, which contains all the regularizing factors. For obtaining the generalized Ward identities we shall use the same method as was used for the proof of the gauge invariance of the S-matrix.

Let us introduce the gauge-invariant function $\bar{\Delta}(\mathcal{A})$ defined by the condition

$$\bar{\Delta}(\mathcal{A}) \int \delta[\partial_\mu \mathcal{A}_\mu^\omega - W(x) - \chi(x)] d\omega = 1, \tag{7.2}$$

where $\chi(x)$ is an arbitrary matrix function. Allowing for (7.2) we rewrite $Z(J)$ as

$$Z(J) = N^{-1} \int \exp\left\{ i \left[S_\Lambda - \frac{1}{2} \operatorname{tr} \int \left[\mathcal{Y}_\mu \mathcal{A}_\mu \right.\right.\right.$$
$$\left.\left.\left. + \frac{1}{2\alpha} (f(\Box) W)^2 \right] dx \right] \right\} \Delta(\mathcal{A}) \bar{\Delta}(\mathcal{A}) \prod_x \delta(\partial_\mu \mathcal{A}_\mu - W)$$
$$\times \delta(\partial_\mu \mathcal{A}_\mu^\omega - W - \chi) d\mathcal{A} dW\, d\omega. \tag{7.3}$$

We pass to new variables

$$\mathcal{A}_\mu \longrightarrow \mathcal{A}_\mu^\omega,$$
$$\omega \longrightarrow \omega^{-1}. \tag{7.4}$$

The integrals over ω and W are removed by δ-functions, and the Jacobian appearing cancels with $\Delta(\mathcal{A})$.

Taking into account that the value of the functional $\tilde{\Delta}(\mathcal{A})$ on the surface,

$$\partial_\mu \mathcal{A}_\mu = W + \chi \tag{7.5}$$

is equal to the value of the functional $\Delta(\mathcal{A})$ on the surface,

$$\partial_\mu \mathcal{A}_\mu = W, \tag{7.6}$$

we obtain

$$Z(J) = N^{-1} \int \exp\left\{ i\left[S_\Lambda - \frac{1}{2}\mathrm{tr} \int \left[\mathcal{Y}_\mu \mathcal{A}_\mu^\omega + \frac{1}{2\alpha}(f(\Box)(\partial_\mu \mathcal{A}_\mu - \chi))^2 \right] dx \right] \right\} \det M \prod_x d\mathcal{A}. \tag{7.7}$$

Here

$$\mathcal{A}_\mu^\omega = \mathcal{A}_\mu + \partial_\mu u - g[\mathcal{A}_\mu, u] + O(u^2), \tag{7.8}$$

and $u(x)$ satisfies the equation

$$\Box u - g\partial_\mu[\mathcal{A}_\mu, u] + O(u^2) = W - \partial_\mu \mathcal{A}_\mu = -\chi. \tag{7.9}$$

Representing the solution u by a series in χ, we have

$$u = -M^{-1}\chi + O(\chi^2), \tag{7.10}$$

where M^{-1} is the inverse operator of M. The kernel of this operator, $M_{ab}^{-1}(x,y)$, satisfies the equation

$$\Box M_{ab}^{-1}(x,y) - gt^{abc}\partial_\mu(\mathcal{A}_\mu^d(x) M_{cb}^{-1}(x,y)) = \delta^{ab}\delta(x-y) \tag{7.11}$$

and obviously coincides with the connected part of the Green function of the fictitious particles in an external classical field $\mathcal{A}_\mu(x)$:

$$M_{ab}^{-1}(x,y) = \delta^{ab} D^0(x-y) + gt^{abc} \int D^0(x-z)\partial_\mu[\mathcal{A}_\mu^d(z) D^0(z-y)]dz + \ldots \tag{7.12}$$

Since the original functional (7.1) does not depend on χ, its derivative with respect to χ is equal to zero:

$$\left.\frac{dZ}{d\chi}\right|_{\chi=0} = 0. \tag{7.13}$$

Substituting into this formula the expression (7.7), obtained by the identity transformation of the functional (7.1), and performing the differentiation explicitly, we get

$$\int \exp\left\{i\left[S_{\mathcal{A}} + \int \left[J_\mu^a \mathcal{A}_\mu^a + \frac{1}{2\alpha}(f(\Box)\partial_\mu \mathcal{A}_\mu)^2\right] dx\right\} \det M$$
$$\times \left\{\frac{1}{\alpha} f^2(\Box)\partial_\mu \mathcal{A}_\mu^a(y)\right.$$
$$\left. + \int J_\mu^b(z)(\nabla_\mu^a M^{-1})^{ba}(z, y, \mathcal{A})dz\right\} \prod_x d\mathcal{A} = 0. \quad (7.14)$$

This equality is nothing but the system of generalized Ward identities for the Yang-Mills theory. It can be written also in terms of variational derivatives

$$\left\{\frac{1}{\alpha} f^2(\Box)\partial_\mu \left\{\frac{1}{i}\frac{\delta}{\delta J_\mu^a(x)}\right\}\right.$$
$$\left. + \int J_\mu^b(y)\left[\nabla_\mu^y\left(\frac{1}{i}\frac{\delta}{\delta J_\mu(y)}\right) M_{yx}^{-1}\left(\frac{1}{i}\frac{\delta}{\delta J}\right)\right]^{ba} dy\right\} Z = 0, \quad (7.15)$$

where the operators $(M^{-1})^{ba}(1/i \cdot \frac{\delta}{\delta J})$ and $\nabla_\mu(1/i \cdot \frac{\delta}{\delta J})$ are derived from $\nabla_\mu(\mathcal{A})$, $(M^{-1})^{ba}(x, y, \mathcal{A})$ by the obvious substitution

$$\mathcal{A}_\mu^a \to \frac{1}{i}\frac{\delta}{\delta J_\mu^a}. \quad (7.16)$$

Applying the operator M^{-1} to $Z(J)$, we obtain the total Green function of the fictitious particles in the presence of the classical source J:

$$(M^{-1})_{xy}^{ab}\left[\frac{1}{i}\frac{\delta}{\delta J}\right] Z(J) = G^{ab}(x, y, J)$$
$$= -\frac{\delta^2}{\delta\eta^{-a}(x)\delta\eta^b(y)} N^{-1} \int \exp\left\{-i\frac{\mathrm{tr}}{2}[\bar{c}\partial_\mu \nabla_\mu c + \mathcal{L}_{YM}\right.$$
$$\left. + \bar{c}\eta + \bar{\eta}c + \mathcal{Y}_\mu \mathcal{A}_\mu]dx\right\} \prod_x d\bar{c}\, dc\, d\mathcal{A}|_{\bar{\eta}=\eta=0} \quad (7.17)$$

This function satisfies the equation

$$\partial_\mu^x \nabla_\mu^{ab}\left(\frac{1}{i}\frac{\delta}{\delta J}\right) G^{bc}(x, y, J) = \delta^{ac}\delta(x - y)Z(J). \quad (7.18)$$

From the generalized Ward identities (7.15) it is easy to obtain relations between various Green functions. For instance, performing variational differentiation of (7.15) with respect to $J_\nu^b(y)$ at $J = 0$ and differentiating the resulting equality with respect to y_ν we get

$$\frac{-i}{\alpha} f^2(\Box)\partial_\mu^x \partial_\nu^y \left[\frac{\delta^2 Z}{\delta J_\mu^a(x)\delta J_\nu^b(y)}\right]_{J=0}$$

$$= -\partial_\nu^y \left\{\nabla_\nu^y \left(\frac{1}{i}\frac{\delta}{\delta J}\right) G_{yx}\right\}_{J=0}^{ba} = -\delta^{ab}\delta(x-y). \quad (7.19)$$

The variational derivative

$$\frac{1}{i}\frac{\delta^2 Z}{\delta J_\mu(x)\delta J_\nu^b(y)}\Bigg|_{J=0} \quad (7.20)$$

is just the two-point Green function of the Yang-Mills field $G_{\mu\nu}^{ab}(x,y)$. The equality (7.19) shows that the longitudinal part of the complete Green function

$$G_{\mu\nu}^L(x-y) = \partial_\mu \partial_\nu \Box^{-2} \partial_\rho \partial_\sigma G_{\rho\sigma}(x-y) \quad (7.21)$$

coincides with the free one,

$$G_{\mu\nu}^L = D_{\mu\nu}^L = -\delta^{ab}\frac{\alpha}{(2\pi)^4}\int e^{-ikx}\frac{k_\mu k_\nu}{k^4 f^2(-k^3)}dk. \quad (7.22)$$

Thus, in complete analogy with electrodynamics, radiative corrections to the longitudinal part of the Green function are absent. The corresponding identities for three- and four-point functions look significantly more complicated than in electrodynamics, since they include in a nontrivial manner the Green functions of fictitious particles.

The consequence of the generalized Ward identities is the existence of relations between counterterms, necessary for the removal of divergences from the Green functions. For example, from the identity (7.19) for the two-point Green function, it follows that the counterterm responsible for the renormalization of the longitudinal part of the wave function is equal to zero. It can be shown that if the Green functions satisfy the generalized Ward identities, then the counterterms form a gauge-invariant structure. This may be done either by directly analyzing the system (7.15) or by passing to analogous identities for one-particle irreducible Green functions. By this it is proved that renormalization does not violate the gauge invariance of the theory. It is simpler, however, to proceed the other way around—from the very beginning to insert in the Lagrangian gauge-invariant counterterms of the most general type, and then, using the generalized Ward identities, to prove that all the Green functions in such a theory tend to a finite limit when the intermediate regularization is removed. That is exactly what we shall do in the next section.

Let us now obtain the generalized Ward identities for the case when the yang-Mills field interacts with scalar φ and spinor ψ fields.

The generating functional for the Green functions in this case can be written as

$$Z(J,\zeta,\bar{\eta},\eta) = N^{-1}\int \exp\Big\{i\Big[S_\Lambda + \int\Big[\frac{1}{2\alpha}(f(\Box)W)^2$$

$$+J^a_\mu A^a_\mu + \zeta^i\varphi^i + \psi^k\eta^k + \bar{\eta}^k\varphi^k\Big]dx\Big]\Big\}$$

$$\times\Delta(A)\prod_x \delta(\partial_\mu A_\mu - W)dAdWd\varphi d\bar{\psi}d\psi. \quad (7.23)$$

The gauge-invariant regularized action also includes terms describing the interaction of spinor and scalar fields.

All the reasoning given above automatically applies to this case. the only difference is that, in addition to the change of variables (7.4), it is necessary to pass to new fields $\varphi^\omega, \psi^\omega$

$$\varphi \to \varphi^\omega, \psi \to \psi^\omega. \quad (7.24)$$

As a result, the additional terms

$$\delta_x(\zeta^i\varphi^i + \bar{\varphi}^k\eta^k + \bar{\eta}^k\psi^k) = \zeta^i\delta_x\varphi^i + \delta_x\psi^k\eta^k + \bar{\eta}^k\delta_x\psi^k \quad (7.25)$$

appear in the exponent in the transformed functional, and the generalized Ward identities take the form

$$\int \exp\Big\{i\Big[S_\Lambda + \int\Big[A^a_\mu J^a_\mu + \zeta^i\varphi^i + \bar{\psi}^j\eta^j + \bar{\eta}^j\psi^j$$

$$+\frac{1}{2\alpha}(f(\Box)\partial_\mu A^a_\mu)^2\Big]dx\Big\}\det M\Big\{\frac{1}{\alpha}f^2(\Box)\partial_\mu A^a_\mu(y)$$

$$+\int\Big[J^b_\mu(z)[\nabla^z_\mu M^{-1}(z,y,A)]^{ba} + \zeta^i(z)\left.\frac{\delta_\chi\varphi^i(z)}{\delta\chi^a(y)}\right|_{\chi=0}$$

$$+\left.\frac{\delta_\chi\bar{\psi}^k(z)}{\delta\chi^a(y)}\right|_{\chi=0}\eta^k(z) + \bar{\eta}^k(z)\left.\frac{\delta_\chi\psi^k(z)}{\delta\chi^a(y)}\right|_{\chi=0}\Big]dz\Big\} = 0. \quad (7.26)$$

In this form the generalized Ward identities are valid both in the symmetric theory and in the theory with spontaneous symmetry breaking. The difference is only in the explicit form of the gauge transformation of the scalar fields φ^ω. If the fields φ realize a representation of the gauge group Ω with the generators Γ^c,

$$\delta\varphi^a = g(\Gamma^c)^{ab}\varphi^b u^c + O(u^2), \quad (7.27)$$

then the identity (7.26) can be written in the form (we omit the spinor fields)

$$\frac{1}{\alpha}f^2(\Box)\partial^x_\mu\left[\frac{1}{i}\frac{\delta Z}{\delta J^a_\mu(x)}\right] + \int\Big[J^b_\mu(y)\Big\{\nabla_\mu\left[\frac{1}{i}\frac{\delta}{\delta J}\right]G(y,x,J,\zeta)\Big\}^{ba}$$

$$+g\zeta^b(y)(\Gamma^d)^{bc}\frac{1}{i}\frac{\delta G^{da}(y,x,J,\zeta)}{\delta\zeta^c(y)}\Big]dy = 0. \quad (7.28)$$

In the case of spontaneously broken symmetry the transformation (7.27) is modified in the following manner:

$$\delta\varphi^a = g(\Gamma^c)^{ab}\varphi^b u^c + g(\Gamma^c)^{ab} r^b u^c + O(u^2), \tag{7.29}$$

where r^b is a constant vector, which may be considered without loss of generality to be directed along the axis labelled $\bar{b}$: $r^b = r\delta^{\bar{b}b}$. Correspondingly there appears in the identity (7.28) an additional term

$$rg \int \zeta^b(y)(\Gamma^d)^{b\bar{b}} G^{da}(y, x, J, \zeta)\,dy. \tag{7.30}$$

For instance, for the model (3.25) of Chapter 1 in which the scalar fields form the complex SU_2 doublet

$$\varphi(x) = \begin{pmatrix} \varphi_1(x) \\ \varphi_2(x) \end{pmatrix} = \frac{1}{\sqrt{2}} \begin{pmatrix} iB_1(x) + B_2(x) \\ \sqrt{2}\mu + \sigma(x) - iB_3(x) \end{pmatrix}, \tag{7.31}$$

the gauge transformation has the form

$$\begin{aligned} \delta\sigma &= \frac{-g}{2}(B^a u^a), \\ \delta B^a &= -m_1 u^a - \frac{g}{2}\varepsilon^{abc} B^b u^c - \frac{g}{2}\sigma u^a, \\ m_1 &= \frac{g\mu}{\sqrt{2}}. \end{aligned} \tag{7.32}$$

The generalized Ward identities appear as follows:

$$\begin{aligned} \frac{1}{\alpha} f^2(\Box)\partial_\mu \left[\frac{1}{i}\frac{\delta Z}{\delta J^a_\mu(x)}\right] &+ \int \Bigg\{ \Bigg[J^b_\mu(y) \nabla^{bd}_\mu \left(\frac{1}{i}\frac{\delta}{\delta J}\right) \\ &- \zeta_\sigma(y) \frac{g}{2}\frac{1}{i}\frac{\delta}{\delta\zeta^d_B(y)} - \zeta^b_B(y)\left(\frac{g}{2}\varepsilon^{bcd}\frac{1}{i}\frac{\delta}{\delta\zeta^c_B(y)} \right. \\ &\left. + \delta^{bd}\frac{g}{2}\frac{1}{i}\frac{\delta}{\delta\zeta_\sigma(y)} + m_1\delta^{bd}\right)\Bigg] G^{da}(y, x, J, \zeta)\Bigg\}\,dy = 0. \end{aligned} \tag{7.33}$$

To conclude this section, we shall show that the generalized Ward identities (7.14) express a certain additional symmetry, having no classical analog, of the effective Lagrangian of the quantized Yang-Mills field. By the effective Lagrangian we mean the expression in the exponent in the formula (3.54) of Chapter 3 for the S-matrix. This expression contains, besides the classical Yang-Mills Lagrangian, a gauge-fixing term and the Lagrangian of fictitious fields.

Let us write the generating functional for the Green functions in the form of a path integral of $\exp\{i \times \text{local action}\}$, introducing explicitly the fields of fictitious particles:

$$Z = N^{-1} \int \exp\Big\{ i \int \Big[-\frac{1}{4} F^a_{\mu\nu} F^a_{\mu\nu} + \frac{1}{2\alpha} (\partial_\mu A_\mu)^2 + \bar{c}^a M^{ab} c^b$$

$$+ J^a_\mu A^a_\mu + \bar{c}^a \eta^a + \bar{\eta}^a c^a \Big] dx \Big\} \prod_x dA d\bar{c}\, dc. \tag{7.34}$$

In this formula, besides the source of for the Yang-Mills field, we have introduced anticommuting sources $\bar{\eta}, \eta$ for the fictitious fields. The functional (7.34) corresponds to a definite choice of the gauge condition; for simplicity we consider the case $f(\Box) = 1$. Therefore the effective action in the exponent is not gauge-invariant. Nevertheless, there exist transformations that affect simultaneously both the Yang-Mills fields and the fictitious fields $\bar{c}, c$ with respect to which the effective Lagrangian is invariant. These transformations have the following form:

$$A^a_\mu(x) \to A^a_\mu(x) + [\nabla_\mu c(x)]^a \varepsilon, \tag{7.35}$$

$$c^a(x) \to c^a(x) - \frac{1}{2} t^{abd} c^b(x) c^d(x) \varepsilon, \tag{7.36}$$

$$\bar{c}^a(x) \to \bar{c}^a(x) + \frac{1}{\alpha} [\partial_\mu A^a_\mu(x)] \varepsilon. \tag{7.37}$$

Here ε is a parameter independent of x and is an element of the Grassman algebra:

$$\varepsilon^2 = 0; \quad \varepsilon c + c\varepsilon = 0; \quad \varepsilon\bar{c} + \bar{c}\varepsilon = 0; \quad [\varepsilon, A_\mu] = 0. \tag{7.38}$$

(We recall that the fictitious fields $\bar{c}, c$ are also anticommuting variables.) Such transformations, which mix up commuting and anticommuting quantities in a non-trivial manner, have become known as supertransformations.

Let us verify that the effective Lagrangian figuring in the formula (7.34) is invariant under the transformations (7.35) to (7.37). The transformation (7.35) is a special case of the gauge transformation, because it leaves the first term in the exponent of (7.34) invariant. It is not difficult to check that the variation $\delta(\nabla_\mu c)$ is also equal to zero. Indeed,

$$\delta(\nabla_\mu c)^a = -\frac{1}{2} t^{abd} [\partial_\mu (c^b c^d) - A^b_\mu t^{def} c^e c^f] \varepsilon - t^{abd} (\partial_\mu c^b - t^{bef} A^e_\mu c^f) \varepsilon c^d. \tag{7.39}$$

The anticommutativity of the variables c^f, c^d leads to

$$t^{abd} t^{bef} c^f c^d = \frac{1}{2} t^{aed} t^{abf} c^b c^f, \tag{7.40}$$

and therefore the right-hand side of (6.39) vanishes. Thus, the total variation of the effective Lagrangian is equal to

$$\delta \mathcal{L}_{ef} = \frac{1}{\alpha} \partial_\mu A^a_\mu \partial_\mu (\nabla_\mu c)^a \varepsilon - \frac{1}{\alpha} \partial_\mu A^a_\mu M^{ab} c^b \varepsilon. \tag{7.41}$$

Remembering the defintion of the operator M, we see that this expression is equal to zero.

Invariance of the effective Lagrangian with respect to the transformations (7.35) to (7.37) can be used for the alternative derivation of the generalized Ward identities (7.15).

For this purpose, in the integral (7.34) we shall make the change of integration variables (7.35) to (7.37). The Jacobian of this transformation is obviously equal to unity. Therefore, as a result of the substitution (7.35) to (7.37) only the terms with sources are changed. Writing out their variation explicity and equating the derivative $dZ/d\varepsilon$ to zero, we obtain the relation

$$\int \exp\left\{i\int[\mathcal{L}_{ef} + J_\mu^a A_\mu^a + \bar{c}^a\eta^a + \bar{\eta}^a c^a]dx\right\}$$
$$\times\left\{J_\mu^b(y)[\nabla_\mu c(y)]^b - \frac{1}{\alpha}\partial_\mu A_\mu^a(y)\eta^a(y) - \right.$$
$$\left. -\frac{1}{2}\bar{\eta}^a(y)t^{abd}c^b(y)c^d(y)\right\} dy \prod_x dA\, d\bar{c}\, dc = 0, \quad (7.42)$$

from which it is easy to derive the generalized Ward identity (7.14). Differentiating the equality (7.42) with respect to η and assuming $\eta, \bar{\eta} = 0$, we have

$$\int \exp\left\{i\int[\mathcal{L}_{ef} + J_\mu^a A_\mu^a]dx\right\}\left\{-\frac{1}{\alpha}\partial_\mu A_\mu^a(y)\right.$$
$$\left. + \int \bar{c}^a(y)J_\mu^b(z)[\nabla_\mu c(z)]^b dx\right\}\prod_x dA\, d\bar{c}\, dc = 0. \quad (7.43)$$

Performing integration over $\bar{c}, c$ by means of the formula (7.17), we obtain exactly the identity (7.14). In an analogous way it is possible to obtain the generalized Ward identities for the case when the Yang-Mills field interacts with the fields of matter.

As a second example of the application of the transformation (7.35) to (7.37), we shall obtain the generalized Ward identities for strongly connected Green functions. Naturally, these identities can be derived directly from the relations (7.15); however, it is more convenient first to pass to the linearized form of the equations (7.15). To this end we introduce into the generating functional (7.34), besides the sources for the fictitious fields $\bar{c}, c$ and the Yang-Mills field $\mathcal{A}_\mu$, sources also for the nonlinear composite fields appearing in the right-hand side of the transformations (7.35) and (7.36).

The modified generating functional will have the form

$$Z(J,\bar{\eta},\eta,K,L) = \int \exp\left\{i\int\left[\mathcal{L}_{ef}(\mathcal{A},\bar{c},c) + J_\mu^a A_\mu^a + \bar{c}^a\eta^a\right.\right.$$
$$\left.\left. + \bar{\eta}^a c^a + K_\mu^a(\nabla_\mu c)^a - \frac{1}{2}L^a t^{abd}c^b d^d\right] dx\right\}\prod_x dA\, d\bar{c}\, dc. \quad (7.44)$$

As in the previous case, we shall change the variables (7.35) to (7.37). The terms containing the sources K and L are invariant with respect to this change of variables. The invariance of the term $\nabla_\mu c$ has been verified above, and the invariance of the product $t^{abd}c^b c^d$ follows directly from the Jacobi identities. Equating $\frac{dZ}{d\epsilon}\big|_{\epsilon=0}$ to zero we obtain

$$\left\{ J_\mu^a(x)\frac{\delta}{\delta K_\mu^a(x)} + \bar{\eta}^a(x)\frac{\delta}{\delta L^a(x)} - \eta^a(x)\frac{1}{\alpha}\left[\partial_\mu \frac{\delta}{\delta J_\mu^a(x)}\right]\right\} Z = 0. \tag{7.45}$$

A completely similar relation is satisfied, obviously, by the generating functional for connected Green functions,

$$W = i \ln Z. \tag{7.46}$$

Strongly connected Green functions are expressed through the variational derivatives of the functional $\Gamma(\mathcal{A}, \bar{c}, \bar{c}, K, L)$ related to the functional W by the Legendre transformation

$$\Gamma(\mathcal{A}, \bar{c}, c, K, L) = -W(J, \eta, \bar{\eta}, K, L) - J_\mu^a \mathcal{A}_\mu^a - \bar{\eta}_a c_a - \bar{c}_a \eta_a, \tag{7.47}$$

where

$$\begin{aligned} \mathcal{A}_\mu^a &= -\frac{\delta W}{\delta J_\mu^a}, \quad J_\mu^a = -\frac{\delta \Gamma}{\delta \mathcal{A}_\mu^a}, \\ c^a &= \frac{\delta W}{\delta \bar{\eta}^a}, \quad \bar{\eta}^a = -\frac{\delta \Gamma}{\delta c^a}, \\ \bar{c}^a &= -\frac{\delta W}{\delta \eta^a}, \quad \eta_n = \frac{\delta \Gamma}{\delta \bar{c}^a}. \end{aligned} \tag{7.48}$$

Indeed, from the definition (7.48) it follows directly that

$$\frac{\delta^2 \Gamma}{\delta \mathcal{A}_\mu^a(x)\delta \mathcal{A}_\nu^b(y)} = -\left[\frac{\delta^2 W}{\delta J_\mu^a(x)\delta J_\nu^b(y)}\right]^{-1} = -G_{\mu\nu}^{-1ab} \tag{7.49}$$

Differentiating (7.49) once again with respect to $\mathcal{A}_\rho^c$ and making using of the definition (7.48), we obtain

$$\begin{aligned} \frac{\delta^3 \Gamma}{\delta \mathcal{A}_\mu^a(x)\delta \mathcal{A}_\nu^b(y)\delta \mathcal{A}_\rho^c(z)} &= -\frac{\delta}{\delta \mathcal{A}_\rho^c(z)} G_{\mu\nu}^{-1ab}(x,y) \\ &= \int G_{\mu\alpha}^{-1ad}(x,\bar{x})\frac{\delta G_{\alpha\beta}^{de}(\bar{x},\bar{y})}{\delta \mathcal{A}_\rho^c(x)} G_{\beta\nu}^{-1eb}(\bar{y},y)d\bar{x}\,d\bar{y} \\ &= \int G_{\mu\alpha}^{-1ad}(x,\bar{x})G_{\nu\beta}^{-1eb}(y,\bar{y})G_{\rho\gamma}^{-1fc}(z,\bar{z}) \\ &\times \frac{\delta^3 W}{\delta J_\alpha^d(\bar{x})\delta J_\beta^e(\bar{y})\delta J_\gamma^f(\bar{z})} d\bar{x}\,d\bar{y}\,d\bar{z}. \end{aligned} \tag{7.50}$$

The equality (7.50) demonstrates that the function $\frac{\delta \Gamma}{\delta \mathcal{A}_\mu^a(x)\delta \mathcal{A}_\nu^b(x)\delta \mathcal{A}_\rho^c(x)}\Big|_{\mathcal{A}=0}$ is obtained from the three-point connected Green function by amputation of the external lines; i.e., it represents a strongly connected Green function. In a totally similar way one can verify that all the variational derivatives of Γ are strongly connected Green functions.

Taking advantage of the definition (7.48), we may rewrite the identity (7.45) in terms of the functional Γ:

$$\frac{\delta\Gamma}{\delta A_\mu^a(x)}\frac{\delta\Gamma}{\delta K_\mu^a(x)}+\frac{\delta\Gamma}{\delta c^a(x)}\frac{\delta\Gamma}{\delta L^a(x)}+\frac{1}{\alpha}\partial_\mu A_\mu^a(x)\frac{\delta\Gamma}{\delta\bar{c}^a(x)}=0. \tag{7.51}$$

The identity (7.51) must be supplemented with an equation for the Green function of the fictitious fields. Performing in the initial generating functional (7.44) a change of variables, $\bar{c}\to\bar{c}+\varepsilon$ and equating the derivative $\frac{dZ}{d\varepsilon}\big|_{\varepsilon=0}$ to zero, we obtain

$$\int(M^{ab}c^b(y)+\eta^a(x))\exp\left\{i\int[\mathcal{L}(\mathcal{A},\bar{c},c)+\ldots]dx\right\}d\mathcal{A}\,d\bar{c}\,dc=0. \tag{7.52}$$

This equation can be rewritten in the form

$$\left(\eta^a(x)+\frac{1}{i}\frac{\partial}{\partial x_\mu}\frac{\delta}{\partial K_\mu^a(x)}\right)Z=0. \tag{7.53}$$

Once again making use of the definition (7.48) we obtain for connected Green functions

$$\frac{\delta\Gamma}{\delta\bar{c}^a(x)}+\frac{\partial}{\partial x_\mu}\frac{\delta\Gamma}{\delta K_\mu^a(x)}=0. \tag{7.54}$$

The equations (7.51) and (7.54) represent generalized Ward identities for strongly connected Green functions. Their characteristic property consists in the absence of any explicit dependence whatsoever on the form of the gauge group. They contain neither any coupling constants nor any structure constants of the symmetry group. The explicit form of the gauge group is utilized only in definition of the sources.

The symmetry with respect to transformations such as (7.35) to (7.37) has turned out to be quite expedient for the investigation of the structure of the effective action. It is especially useful in gauge theories with open algebra, i.e., in theories in which the algebra of the gauge group closes only on the solutions of the equations of motion. An example of such theories is presented by the models of extended supergravity.

This symmetry also serves as the basis of the explicitly covariant approach to the quantization of gauge theories, which is the analog of the Gupta-Bleuler formalism in quantum electrodynamics. This approach will be described in one of the sections below.

Until now we have considered only covariant α-gauges, which are usually dealt with in practical calculations. However, all the reasoning automatically applies to the more general case when the term fixing the gauge has the following form:

$$B(\mathcal{A})\exp\left\{\frac{i}{2}\int\Phi^2(\mathcal{A},x)dx\right\}, \tag{7.55}$$

where $\Phi(\mathcal{A})$ i a functional of $\mathcal{A}(x)$, which in principle can involve, in addition to terms linear in $\mathcal{A}$, terms of higher order as well. In this case, in accordance with the general quantization procedure described in Chapter 3, the operator M figuring in the generating functional for the Green functions is given by the formula (1.26) of Chapter 1:

$$M_\Phi \alpha = \int \frac{\delta \Phi(\mathcal{A}, x)}{\delta \mathcal{A}_\mu(y)} \nabla_\mu \alpha(y) dy. \tag{7.56}$$

To obtain the generalized Ward identities of this case it is sufficient, in all the computations given at the beginning of the present section, to substitute

$$\partial_\mu \mathcal{A}_\mu \to \Phi(\mathcal{A}), \qquad M \to M_\Phi. \tag{7.57}$$

As a result, instead of (6.14) we obtain

$$\int \exp\left\{ i \int \left\{ S_\Lambda + \left[\frac{1}{2} \Phi^2(\mathcal{A}) + J^a_\mu A^a_\mu \right] dx \right\} \det M_\Phi \right.$$
$$\times \left\{ \Phi^a(\mathcal{A}, y) + \int J^b_\mu(z) (\nabla^z_\mu M^{-1}_\Phi)^{ba}(z, y, \mathcal{A}) dz \right\} \prod_x d\mathcal{A} = 0. \tag{7.58}$$

4.8 The Structure of the Renormalized Action

Let us analyze the structure of the primitively divergent diagrams in the Yang-Mills theory. We shall start with the Yang-Mills field in vacuum. In the α-gauge the effective Lagrangian has the form

$$\mathcal{L} = \frac{1}{2} \operatorname{tr} \left\{ \frac{1}{4} [(\partial_\nu \mathcal{A}_\mu - \partial_\mu \mathcal{A}_\nu) + g[\mathcal{A}_\mu, \mathcal{A}_\nu]]^2 \right.$$
$$\left. - \frac{1}{2\alpha} (f(\Box) \partial_\mu \mathcal{A}_\mu)^2 + \bar{c} \Box c - g \bar{c} \partial_\mu [\mathcal{A}_\mu, c] \right\}. \tag{8.1}$$

The diagram technique involves the following elements:

1. Vector lines $\overline{\mathcal{A}_\mu \mathcal{A}_\nu}$, lines of fictitious c particles $\overline{\bar{c}c}$. To these lines correspond the free Green functions $D_{\mu\nu}(p)$ and $D(p)$, behaving asymptotically as p^{-2}, as $p \to \infty$.
2. Vertices with three outgoing vector lines and one derivative.
3. Vertices with four vector lines and without derivatives.
4. Vertices with a single vector and two fictitious lines and one derivative.

In accordance with the general formula derived in Section 4.2, the index of a diagram containing n_3 three-legged vector vertices, n_4 four-legged vertices, n_c vertices with fictitious particles participating, L_{in} internal vector lines, and L^c_{in} internal fictitious lines is equal to

$$\omega = 2L_{\text{in}} + 3L^c_{\text{in}} - 4(n_4 - 1) - 3(n_3 + n_c). \tag{8.2}$$

Taking advantage of the fact that the number of internal lines L_{in}, L^c_{in} is related to the number of external lines L_{ex} by

$$L_{in} = \frac{4n_4 + 3n_3 + n_c - L_{ex}}{2}, \qquad L^c_{in} = \frac{2n_c - L^c_{ex}}{2} \tag{8.3}$$

we express the diagram index in terms of the number of external lines:

$$\omega = 4 - L_{ex} - L^c_{ex}. \tag{8.4}$$

From this formula it follows that the perturbation-theory series for the Yang-Mills field in the α-gauge contains a finite number of types of primitively divergent diagrams. These diagrams are symbolically presented in Fig. 18. Formally, a logarithmically divergent diagram also exists with two external vectors and two fictitious lines; and a divergent diagram exists which has four fictitious lines. However, as is seen from the formula (8.1), the derivative in one of the vertices can be transposed to

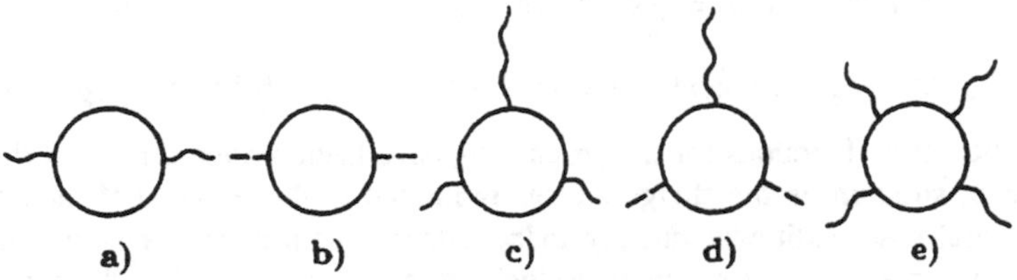

Figure 18. Types of divergent diagrams in the Yang-Mills theory.

the external c-line by integrating by parts. Therefore in reality the corresponding diagrams converge.

The self-energy diagrams in (a) and (b) have an index 2. The index of the diagrams in (c) and (d) is 1, and of the diagram in (e), 0. For the same reasons as above, the actual index of diagrams having external c-lines is reduced by 1. Besides this, for reasons of Lorentz invariance all diagrams with indices equal to 1 in reality diverge only logarithmically.

In accordance with the general procedure for the removal of these divergences, it is necessary to subtract from the corresponding vertex functions several of the leading terms of the Taylor series in the external momenta. As the expansion center, a point is usually chosen at which the external momenta are on the mass shell, since such a choice provides for the proper normalization of one-particle states. However, in the case when the system under consideration involves particles of zero mass, the vertex functions on the mass shell may involve additional singularities due to the divergence of the corresponding integrals at the origin of the coordinate system (the infrared catastrophe). Therefore we shall perform subtractions at the points for which the values of all the external momenta are in the Euclidean region. For a vertex with n external momenta p_i, for example, such points are

$$p_i^2 = -a^2, \quad p_i p_j = \frac{a^2}{n-1}. \tag{8.5}$$

At these points all the vertex functions are real and are free from infrared singularities.

We shall now write out the most general expressions for the subtracted terms compatible with the conditions of relativistic invariance and of Bose symmetry. The proper vertex functions corresponding to the diagrams presented in Fig. 18 have the following structure:

$$\begin{aligned}
\Gamma_{AA}(p) &\equiv \Gamma^{ab}_{\mu\nu}(p) = \delta^{ab}\{b_1 g_{\mu\nu} + b_2 p_\mu p_\nu + b_3(p^2 g_{\mu\nu} - p_\mu p_\nu)\} + \ldots, \\
\Gamma_{\bar{c}c}(p) &\equiv \Gamma^{ab}(p) = \delta^{ab} b_4 p^2 + \ldots, \\
\Gamma_{A^3} &\equiv \Gamma^{lmn}_{\lambda\nu\rho}(p,k,q) = \hat{i}\varepsilon^{lmn} b_5 \{g_{\lambda\nu}(p-k)_\rho \\
&\qquad + g_{\nu\rho}(k-q)_\lambda + g_{\lambda\rho}(q-p)_\nu\} + \ldots, \\
\Gamma_{A\bar{c}c} &\equiv \Gamma^{lmn}_\mu(p,k,q) = \frac{i}{2}\varepsilon^{lmn} b_6 (k-q)_\mu + \ldots, \\
\Gamma_{A^4} &\equiv \Gamma^{ilmn}_{\mu\nu\rho\lambda}(p,q,k,r) = P\{b_7 \varepsilon^{gil}\varepsilon^{gmn} g_{\mu\rho} g_{\nu\lambda} + b_8 \delta^{il}\delta^{mn} g_{\mu\nu} g_{\lambda\rho}) + \ldots
\end{aligned} \tag{8.6}$$

(Here the vertex functions for the group SU_2 have been written down. In this case the tensor structure in the charge indices is limited to the tensors ε^{abc} and δ^{ab}. In the general case additional linearly independent structures may be present, such as, for example, terms proportional to the symmetric tensor d^{abc} in the case of the group SU_3. The proof of the renormalizability given below remains valid in this case too.)

In the last formula, P is the symmetrization operator of the pairs of indices $(i,\mu), (1,\nu), (m,\rho), (n,\lambda)$. The coefficients b_i depend on the regularization parameters Λ, μ and on the positions of the subtraction points $a_i, \ldots$ and the ellipsis denotes the subsequent terms of the Taylor expansion, which tend to a finite limit when the regularization is removed. The subtraction of the polynomials (8.6) is equivalent to the insertion in the effective Lagrangian of the following counterterms:

$$\begin{aligned}
\Delta\mathcal{L} = \frac{1}{4}\mathrm{tr}\Big\{ & b_1 A_\mu^2 + b_2(\partial_\mu A_\mu)^2 + \frac{b_3}{2}(\partial_\nu A_\mu - \partial_\mu A_\nu)^2 \\
& + 2b_4 \bar{c}\Box c + b_5(\partial_\nu A_\mu - \partial_\mu A_\nu)[A_\mu, A_\nu] + \\
& + 2b_6 \bar{c}\partial_\mu [A_\mu, \bar{c}] + \frac{b_7}{2}[A_\mu, A_\nu][A_\mu, A_\nu]\Big\} + \frac{b_8}{16}\{\mathrm{tr}(A_\mu, A_\mu)\}^2.
\end{aligned} \tag{8.7}$$

The number of types of counterterms needed for the removal of divergences is finite, and therefore the theory is renormalizable. However, the counterterm part of the action (8.7) contains significantly more parameters than the original Lagrangian.

For arbitrary parameters b_i the renormalized theory is not gauge-invariant and does not satisfy the relativity principle, and that leads to the loss of the equivalence of various gauges and as a consequence to the violation of unitarity.

When the intermediate regularization is fixed, then the values of the parameters b_i depend on the choice of the subtraction points a_i. We shall show that this arbitrariness allows one to choose the parameters b_i so that the renormalized theory becomes gauge-invariant.

Let us ascertain what constraints are imposed by the relativity principle on the form of the renormalized effective Lagrangian. First of all note that the gauge-invariance of the Yang-Mills Lagrangian is not violated if it is multiplied by a constant. In addition we can treat arbitrarily the parameter g, which plays the part of the charge. In other words, the most general expression for the gauge-invariant Lagrangian of the Yang-Mills field has the form

$$\mathcal{L} = \frac{z_2}{8}\operatorname{tr}\{(\partial_\nu \mathcal{A}_\mu - \partial_\mu \mathcal{A}_\nu) + g z_1 z_2^{-1}[\mathcal{A}_\mu, \mathcal{A}_\nu]\}^2. \tag{8.8}$$

The role of the parameter of the gauge transformation for this Lagrangian is played by the constant $\bar{g}$:

$$\bar{g} = g z_1 z_2^{-1}. \tag{8.9}$$

The same parameter must be involved in the definition of the covariant derivative to provide for the self-consistency of the theory. Specifically, in the operator M, which has been defined by the formula

$$M = \partial_\mu \nabla_\mu = \partial_\mu(\partial_\mu - g[\mathcal{A}_\mu, \quad]), \tag{8.10}$$

the constant $\bar{g}$ must be substituted for g. Writing $\det M(\bar{g})$ in the form of an integral over the fields of fictitious particles,

$$\det M(\bar{g}) = \text{const} \int \exp\left\{\frac{-i\operatorname{tr}}{2}\int \tilde{z}_2 \bar{c}\partial_\mu(\partial_\mu c - \tilde{z}_1 \tilde{z}_2^{-1} g \times [\mathcal{A}_\mu, c])dx\right\} \prod_x d\bar{c}\, dc, \tag{8.11}$$

where

$$\tilde{z}_2^{-1}\tilde{z}_1 = z_2^{-1} z_1 \tag{8.12}$$

and we have again taken the opportunity to multiply the Lagrangian as a whole by an arbitrary constant, we obtain the most general expression for the admissible effective Lagrangian:

$$\begin{aligned}\mathcal{L}_R = \frac{1}{2}\operatorname{tr}\Big\{\frac{1}{4} z_2[(\partial_\nu \mathcal{A}_\mu - \partial_\mu \mathcal{A}_\nu) + z_1 z_2^{-1} g[\mathcal{A}_\mu, \mathcal{A}_\nu]]^2 \\ -\frac{1}{2\alpha}[f(\Box)\partial_\mu \mathcal{A}_\mu]^2 - \tilde{z}_2(\bar{c}\Box c - \tilde{z}_1 \tilde{z}_2^{-1} g \bar{c}\partial_\mu[\mathcal{A}_\mu, c])\Big\}. \end{aligned} \tag{8.13}$$

The constants $z_1, z_2, \tilde{z}_1, \tilde{z}_2$ are related by (8.12). The familiar condition $z_1 = z_2$ is not necessary and, generally speaking, does not hold: The Lagrangian (8.13) has the same structure as the nonrenormalized Lagrangian, differing from the latter only by the multiplicative renormalization of the fields $\mathcal{A}_\mu, \bar{c}, c$ of the charge g, and of the gauge parameter α:

$$\begin{gathered} \mathcal{A}_\mu \to z_2^{1/2}\mathcal{A}_\mu, \quad c \to \tilde{z}_2^{1/2}c, \quad \bar{c} \to \tilde{z}_2^{1/2}\bar{c}, \\ g \to z_1 z_2^{-3/2} g, \quad \alpha \to z_2\alpha. \end{gathered} \tag{8.14}$$

Unlike the general expression (7.7), this Lagrangian involves only three independent counterterms, and at first sight it is not obvious that with their aid it is possible to remove all the divergences.

By introducing an invariant intermediate regularization it is possible to construct a generating functional for the Green functions $Z_R(J)$ corresponding to the Lagrangian (8.13). We shall assume that the regularization is performed, using the method of higher covariant derivatives described in Section 4.4. Since in all further reasoning only the invariance of the regularized action will be used, we shall not write the regularizing terms explicitly; they are described, for example, by the formula (4.26).

The role of the parameter of the gauge transformation for the Lagrangian (8.13) is played by the constant $\tilde{g} = z_1 z_2^{-1} g$. Therefore, the generalized Ward identities that are satisfied by the functional $Z_R(J)$ differ from the identities (7.15) by the substitution of $\tilde{g}$ for g in the operators M^{-1} and ∇_μ:

$$\begin{aligned} &\frac{1}{\alpha} f^2(\Box)\partial_\mu^x \left\{ \frac{1}{i}\frac{\delta Z_R}{\delta J_\mu^a(x)} \right\} \\ &\qquad + \left\{ \int J_\mu^b(y) \left[\delta^{bd}\partial_\mu^y - \tilde{g} t^{bcd} \frac{1}{i}\frac{\delta}{\delta J_\mu^c(y)} \right] \right. \\ &\qquad\qquad \left. \times \tilde{M}_{yx}^{-1da} \left(\frac{1}{i}\frac{\delta}{\delta J} \right) dy \right\} Z_R = 0, \end{aligned} \tag{8.15}$$

where the sign $\tilde{}$ means that the constant g involved in the definition of the operator M^{-1} is replaced by $\tilde{g}$.

In this formula it is convenient to pass to the renormalized Green functions of the fictitious particles, defined by the equality

$$\begin{aligned} G_R^{da}(y,x,J) = N^{-1} \int \bar{c}^d(y)c^a(x)\exp\left\{ i \int [\tilde{z}_2\bar{c}^i(s)\Box c^i(s) \right. \\ \left. - \tilde{z}_1 g^{ikl} g\bar{c}^i(s)\partial_\mu[A_\mu^k(s)c^l(s)] + \mathcal{L}_R^{YM}(s) + A_\mu^a J_\mu^a]ds \right\} \prod_x dA\, d\bar{c}\, dc. \end{aligned} \tag{8.16}$$

For this, note that

$$\tilde{M}_{yx}^{-1ab} Z_R = N^{-1} \int \bar{c}^a(y) c^b(x) \exp\Big\{ i \int [\bar{c}^a(s) \Box c^a(s) - \tilde{z}_2^{-1} \tilde{z}_1 g t^{abd} \bar{c}^a(s) \partial_\mu [A_\mu^b(s) c^d(s)] + \mathcal{L}_R^{YM}(s) + A_\mu^a J_\mu^a] ds \Big\} \prod_x dA\, d\bar{c}\, dc = \tilde{z}_2 G_R^{ab}(y, x, J). \tag{8.17}$$

The last relation is obtained from (8.16) by the substitution of variables

$$c \to \tilde{z}_2^{-1/2} c, \quad \bar{c} \to \tilde{z}_2^{-1/2} \bar{c}. \tag{8.18}$$

Representing the source J_μ on the right-hand side of the identity (8.17) in the form $J_\mu = J_\mu^{tr} + \partial_\mu \Box^{-1} \partial_\nu J_\nu$, and taking advantage of the fact that

$$\partial_\mu^y (\nabla_\mu(\tilde{g}) \tilde{M}^{-1})_{yx}^{ab} = \delta^{ab} \delta(x - y), \tag{8.19}$$

we rewrite these relations in the form

$$\frac{1}{\alpha} f^2(\Box) \partial_\mu^x \left\{ \frac{1}{i} \frac{\delta Z_R}{\delta J_\mu^a(x)} \right\} = \left\{ \int D_0(x - y) \partial_\mu J_\mu^a(y) dy \right\} Z_R + \int J_\mu^{trb}(y) g \tilde{z}_1 t^{bcd} \frac{1}{i} \frac{\delta}{\delta J_\mu^c(y)} G_R^{da}(y, x, J) dy. \tag{8.20}$$

We shall show that for a suitable choice of the constants $z_2, \tilde{z}_2, \tilde{z}_1$, the finiteness of all the Green functions follows from the identity (8.20). The proof will be by induction. Assuming all the diagrams up to and including the n-th order to be finite, we shall show that the functional F on the right-hand-side of the equation (8.20) is finite to the order $n + 1$. Hence it follows that the functional

$$\partial_\mu^x \left\{ \frac{1}{i} \frac{\delta Z_R}{\delta J_\mu^a(x)} \right\} \tag{8.21}$$

is also finite to order $n+1$. This, as we shall see, means that all the Green functions (except, maybe, the two-point functions of the Yang-Mills fields Γ_{AA} and of the fictitious particles $\Gamma_{\bar{c}c}$) are finite. The divergences of these functions are removed by using the renormalization constants z_2 and $\tilde{z}_2$, the choice of which remains at our disposal.

The proof is particularly simple in the case of gauges for which the longitudinal part of the Green function of the Yang-Mills field decreases rapidly at large momenta, that is, when the function $f(k^2)$ involved in the definition of the generalized α-gauge behaves asymptotically as k^{2n} for $n > 1$. This, of course, includes also the Lorentz gauge itself ($\alpha = 0$) when the longitudinal part of $D_{\mu\nu}$ is equal to zero.

The proof for an arbitrary gauge does not involve any notions that are new in principle, but it is more cumbersome. So, not to distract the reader with technical details, we shall first consider the case and come back to the discussion of more general gauges later, when we are investigating the dependence of the Green functions on the gauge.

In the lowest order of g^2 only the two-point functions Γ_{AA} and $\Gamma_{\bar{c}c}$ diverge. The proper vertex function $\Gamma_{AA} \equiv \Gamma_{\mu\nu}^{m}(k)$ in second order is connected with the Green function $G_{\mu\nu}^{ab}(k)$ by the relation

$$G_{\alpha\beta}^{ab}(k) = D_{\alpha\mu}^{am}(k)\Gamma_{\mu\nu}^{mn}(k)D_{\nu\beta}^{nb}(k) + D_{\alpha\beta}^{ab}(k). \tag{8.22}$$

From the identity (8.20) it follows that

$$\frac{f^2(-k^2)}{\alpha}k^\alpha k^\beta G_{\alpha\beta}^{ab}(k) = \frac{\alpha k_\mu}{k^2}\Gamma_{\mu\nu}^{ab}(k^2)\frac{k_\nu}{k^2 f^2(-k^2)} + \delta^{ab} = \delta^{ab}, \tag{8.23}$$

that is, the function $\Gamma_{\mu\nu}^{ab}(k)$ is transverse:

$$\Gamma_{\mu\nu}^{ab}(k) = \delta^{ab}(k^2 g_{\mu\nu}) - k_\mu k_\nu)\Pi(k^2). \tag{8.24}$$

Therefore the constants b_1 and b_2 in the Lagrangian (8.7) are equal to zero, and for the removal of the divergence one counterterm $z_2^{(0)}$ is sufficient. In an analogous way the counterterm $\tilde{z}_2^{(0)}$ provides for the finiteness of the Green function of the fictitious particles.

We shall now prove the finiteness of the third-order vertex functions. To the vertex functions $\Gamma_{\bar{c}cA}$ there correspond the strongly connected diagrams presented in Fig. 19. As it was shown above, the index of these diagrams equals zero, which means they are formally logarithmically divergent. It is not difficult to verify, however, that in the gauges for which $f(k^2) \underset{k\to\infty}{\to} |k|^{2n}, n \geq 1$, the divergence is absent. Indeed, the analytical expression corresponding to the diagrams in Fig. 19, represents in the momentum representation the sum of terms of the type

$$-\int D(x - x_1)\partial_\nu^{x_1}\{D(x_1 - z_1)\partial_\rho^{z_1}[D_{\rho\mu}(z_1 - z)D(z_1 - y_1)]$$
$$\times \partial_\lambda^{y_1}[D_{\nu\lambda}(x_1 - y_1)D(y_1 - y)]\}dx_1\, dy_1\, dz_1.$$

Integrating by parts, it is easy to transform this expression so that the derivatives at the vertices x_1 and y_1 act either on the vector Green function, as a result of which only its rapidly decreasing longitudinal part gives a contribution to the integral, or on the external line. Therefore the true index of divergence is reduced, and convergent integrals correspond to the diagrams in Fig. 19. Correspondingly, in the gauges under consideration the constant z_1 is finite.

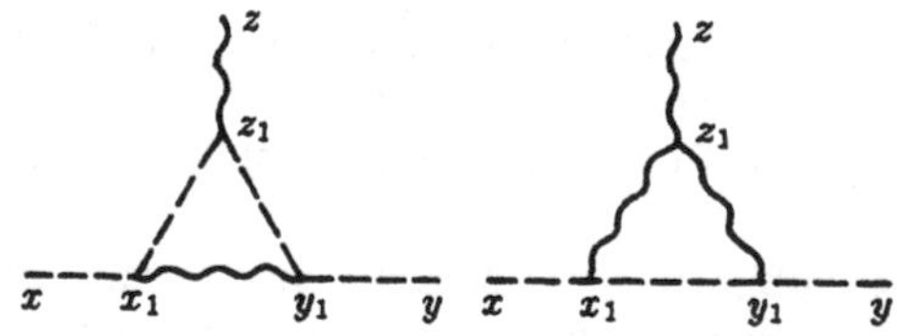

Figure 19. Third-order corrections to the vertex function $\Gamma_{\bar{c}cA}$.

For the Green function $G^{abc}_{\mu\nu\rho}(x,y,z)$ the identity (8.20) gives

$$\frac{1}{\alpha}f^2(\Box)\partial^x_\mu G^{abc}_{\mu\nu\rho}(x,y,z) = g\tilde{z}_1 t^{bed}\left(g_{\nu\alpha} - \frac{\partial_\nu\partial_\alpha}{\Box}\right)$$

$$\times\frac{1}{(i^3)}\left.\frac{\delta^2 G^{da}_R(y,x,J)}{\delta J^e_\alpha(y)\delta J^c_\rho(z)}\right|_{J=0} + (b \leftrightarrow c, y \leftrightarrow z, \nu \leftrightarrow \rho). \quad (8.25)$$

The diagrams in Fig. 20 correspond to the function $\left.\frac{\delta^2 G^{da}_R(y,x,J)}{\delta J^e_\alpha(y)\delta J^c_\rho(Z)}\right|_{J=0}$. The diagrams (b) and (c) are weakly connected, and the convergence of the corresponding integrals follows from the finiteness ofthe second-order two point green functions $G^{ab}_{\mu\nu}$ and G^{ab}.

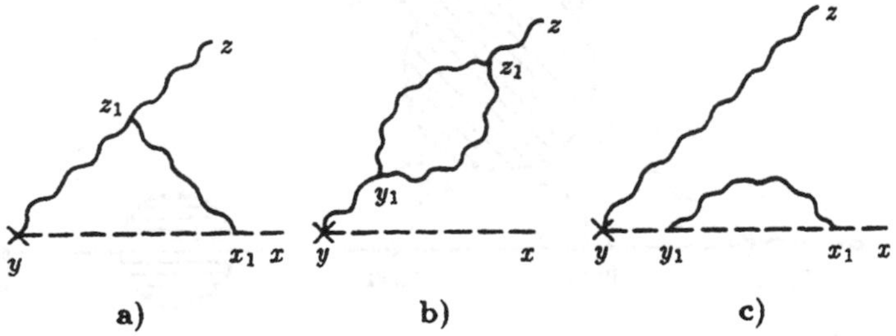

Figure 20. Diagrams corresponding to the function $\left.\frac{\delta^2 G^{da}_R(y,x,J)}{\delta J^e_\alpha(y)\delta J^c_\rho(z)}\right|_{J=0}$.

The diagram in Fig. 20(a) has a structure analogous to the structure of the diagram describing the transition of two fictitious particles into one vector particle (see Fig. 19). It differs from the latter only in the form of the extreme left-hand vertex, denoted in the figure by a small cross. Only one vector and one fictitious line leave this vertex, and derivatives are absent. Formally the diagram in Fig. 20, a has a zero index, but for the same reasons as above the actual index is smaller by unity, and the divergence is absent.

Thus, the right-hand side of the equality (8.25) is finite, and consequently the function $\partial^x_\mu G^{abc}_{\mu\nu\rho}(x,y,x)$ is also finite in the third order of g. Therefore the divergence of the upper vertex function $\Gamma^{abc}_{\mu\nu\rho}$ is also finite:

$$(k+p)_\mu \Gamma^{abc}_{\mu\nu\rho}(k,p) < \infty. \quad (8.26)$$

The divergent part of $\Gamma^{abc}_{\mu\nu\rho}(k,p)$ can only be a polynomial of order not higher than one. The condition (8.26) signifies that this polynomial is identically equal to zero, and consequently, the function $\Gamma^{abc}_{\mu\nu\rho}(k,p)$ is finite in the third order of g. For the removal of the divergence from the vertex function $\Gamma^{abc}_{\mu\nu\rho}$ we have not had to introduce the independent renormalization constant z_1. This function turns out to be finite automatically if $z_1 = \tilde{z}_1\tilde{z}_2^{-1}z_2$.

The proof of the finiteness of the vertex functions of arbitrary order is absolutely analogous.

Let us consider the functional F in the right-hand side of equation (8.20). Its connected part is represented by the diagrams in Fig. 21. Let all the diagrams involved in the expansion of this functional be finite in all orders up to n. To prove the finiteness of the functional F in the $(n+1)$th order, it is sufficient to consider the diagrams in Fig. 21 in which insertions in the external lines are absent, all the subgraphs being assumed to be finite.

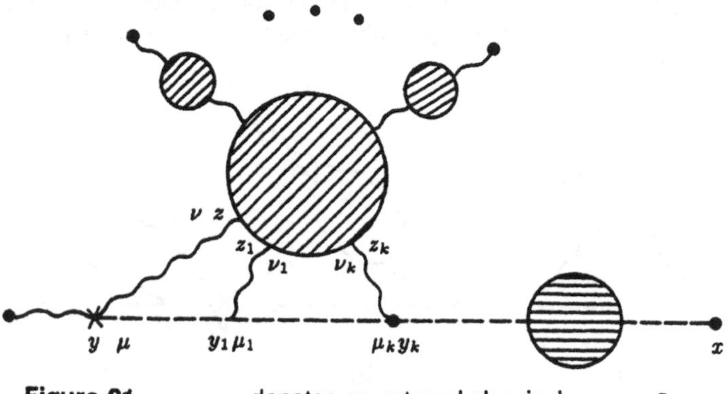

Figure 21. $\sim\!\sim$ denotes an external classical source J_μ.

The diagrams in Fig. 21 are analogous to the diagrams representing Green functions $G(x, y, J)$ with two external fictitious lines and an arbitrary number of external vector lines. They differ only in the form of the vertex denoted by a small cross. One vector and one fictitious line leave this vertex, and derivatives are absent.

Therefore, the index of the diagrams in Fig. 21 is the same as that of the diagrams corresponding to $G(x, y, J)$. The diagram with two external lines has an index equal to one. For reasons of Lorentz invariance, the corresponding analytical expression has the form

$$\int J_\mu^{\mathrm{tr}}(y)\Phi_\mu(y-x)dy = \int J_\mu^{\mathrm{tr}}(y)\partial_\mu\Phi(y-x)dy = 0.$$

The diagram with three external lines has a zero index and in principle diverges logarithmically. The remaining diagrams all have negative indices. Repeating word for word the reasoning given above for the third-order diagrams, we come to the conclusion that logarithmic divergences in the diagrams in Fig. 21 and also in the diagrams responsible for the vertex function $\Gamma_{\bar{c}cA}$ are indeed absent.

Thus the assumption that the integrals corresponding to diagrams of the n-th order are convergent leads to the finiteness of the functional F in all orders up to $n+1$ inclusively. This means that all the integrals involved in the expansion of the functional in the left-hand side of the equation (8.20) also converge, and thus all the Green functions are finite:

$$\partial^{x_1}_{\mu_1}\left\{\frac{1}{i}\frac{\delta}{\delta J^{a_1}_{\mu_1}(x_1)}\cdots\frac{1}{i}\frac{\delta}{\delta J^{a_m}_{\mu_m}(x_m)}\right\}Z_R = \partial^{x_1}_{\mu_1}G^{a_1\ldots a_m}_{\mu_1\ldots\mu_m}(x_1\ldots x_m). \tag{8.27}$$

The Fourier transforms of the function $G^{a_1\ldots a_m}_{\mu_1\ldots\mu_m}$ are related to the vertex functions $\Gamma^{a_1\ldots a_m}_{\mu_1\ldots\mu_m}$ by the relation

$$G^{a_1\ldots a_m}_{\mu_1\ldots\mu_m}(k_1\ldots k_m) = G^{a_1b_1}_{\mu_1\nu_1}(k_1)\ldots G^{a_mb_m}_{\mu_m\nu_m}(k_m)\Gamma^{b_1\ldots b_m}_{\nu_1\ldots\nu_m}(k_1\ldots k_m). \tag{8.28}$$

All the two-point functions $G^{a_ib_i}_{\mu_i\nu_i}$ are reversible and are of order $\leq n$ in g. Therefore from the finiteness of the functions (8.27) it follows that

$$k_{\mu_1}\Gamma^{a_1\ldots a_m}_{\mu_1\ldots\mu_m}(k_1\ldots k_m) = \varphi(k_1\ldots k_m) < \infty. \tag{8.29}$$

Generally speaking, both strongly and weakly connected diagrams correspond to the vertex functions $\Gamma^{a_2\ldots a_m}_{\mu_1\ldots\mu_m}$. To weakly connected diagrams there correspond coefficient functions representable (in the momentum representation) in the form of a product of coefficient functions of a lower order, which are assumed to be finite. Therefore, the equality (8.29) can be considered to be correct for the proper vertex functions of the order $n+1$.

If the proper vertex function has an index 0 or 1, then its divergent part can only be a polynomial of order not higher than one. The condition (8.29) signifies that this polynomial is identically equal to zero, and consequently, all the functions $\Gamma^{a_1\ldots a_m}_{\mu_1\ldots\mu_m}(k_i\ldots k_m)$ are finite. The only possible exceptions are the two-point Green functions of the Yang-Mills field and of the field of the fictitious c-particles. The latter is not even involved in the expansion (8.27) and therefore is not subject to any limitations.

With regard to the two-point Green function of the Yang-Mills fields, since the diagram corresponding to it has an index 2, the divergent part of $\Gamma^{a_1a_2}_{\mu_1\mu_2}(k)$ can be a second-order polynomial. The condition (8.29) is insufficient for turning a second-order polynomial into zero. It does not impose any restrictions on its transverse part,

$$\text{const}\,(g_{\mu\nu}k^2 - k_\mu k_\nu). \tag{8.30}$$

We have, however, at our disposal two more arbitrary counterterms of order $n+1$: $z_2^{(n+1)}, \tilde{z}_2^{(n+1)}$. These counterterms are sufficient for the removal of divergences from the two-point Green functions. Thus, all the diagrams of the $(n+1)$th order are finite. The induction has come to an end.

Now let the Yang-Mills field interact also with scalar and spinor fields. The corresponding Lagrangians are given by the formulas (1.3.1, 11). The diagram technique, besides the elements already discussed, contains now scalar and spinor lines, to which there correspond the Green functions $D(p)$ and $S(p)$ with the asymptotic behavior p^{-2} and p^{-1}, respectively, vertices with two spinor lines and one vector line without derivatives, vertices with two scalar lines and one vector line with one derivative, and vertices with two vector and two scalar lines without derivatives. The index of the diagram with L^A_{ex} external vector, L^c_{ex} fictitious, L^φ_{ex}

scalar, and L^{ψ}_{ex} spinor external lines is equal to

$$\omega = 4 - L^A_{ex} - L^c_{ex} - L^{\varphi}_{ex} - \frac{3}{2}L^{\psi}_{ex}. \tag{8.31}$$

Besides the diagrams already mentioned, the diagrams presented in Fig. 22 also have a nonnegative index. The self-energy diagram for the scalar field in (*c*) diverges quadratically, the diagrams in (*a*), (*d*) linearly, and the remaining diagrams logarithmically.

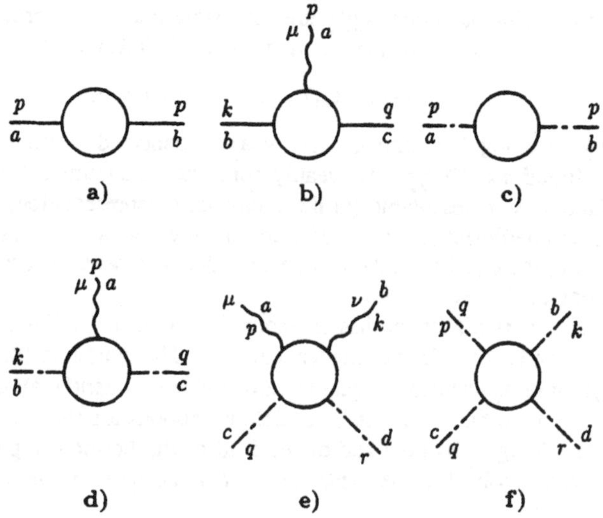

Figure 22. Additional divergent diagrams in the theory of the Yang-Mills field interacting with spinor and scalar fields. The solid line denotes the propagator of a spinor particle; the dash-dotted line that of a scalar particle.

The corresponding proper vertex functions have the form (again, for definiteness, we write the formulas for the case of the group $SU(2)$)

$$\begin{aligned}
\Gamma_{\bar{\psi}\psi} &= (d_1 + \gamma_\mu p_\mu d_2)\delta^{ab} + \dots,\\
\Gamma_{\varphi\varphi} &= (d_3 + d_4 p^2)\delta^{ab} + \dots,\\
\Gamma_{\bar{\psi}\psi A} &= d_5 \gamma_\mu \varepsilon^{abc} + \dots,\\
\Gamma_{\varphi\varphi A} &= i d_6 \varepsilon^{abc}(k-q)^\mu + \dots,\\
\Gamma_{\varphi\varphi AA} &= d_7 g_{\mu\nu}(\delta^{ab}\delta^{cd} - \delta^{ac}\delta^{bd}) + d_8 g_{\mu\nu}(\delta^{ab}\delta^{cd} + \delta^{ac}\delta^{bd}) + \dots,\\
\Gamma_{\varphi^4} &= d_9(\delta^{ab}\delta^{cd} + \delta^{ac}\delta^{bd} + \delta^{ad}\delta^{bc}) + \dots,
\end{aligned} \tag{8.32}$$

where ... stands for the terms tending to a definite limit when the regularization is removed. As in the case of the Yang-Mills field in vacuum, the number of possible counterterms is significantly greater than the number of parameters in the nonrenormalized Lagrangians.

The most general expression for the gauge-invariant renormalized Lagrangian is constructed as before, and has the form

$$\mathcal{L}^R = L^R_{YM} + iz_{2\psi}\bar{\psi}\gamma_\mu(\partial_\mu - z_{2\psi}^{-1}z_{1\psi}gA^a_\mu\Gamma(T^a))\psi$$
$$-z_{2\psi}(m+d)\bar{\psi}\psi + \frac{1}{2}z_{2\varphi}(\partial_\mu\varphi - z_{1\varphi}z_{2\varphi}^{-1}gA^a_\mu\Gamma(T^a)\varphi)^2$$
$$-\frac{z_{2\varphi}}{2}(m^2+f)\varphi^2 + z\lambda(\varphi^2)^2, \qquad (8.33)$$

where $\mathcal{L}^R_{YM}$ is the renormalized Lagrangian of the Yang-Mills field in vacuum (8.13).

Gauge invariance requires that the constants $z_{2\psi}^{-1}z_{1\psi}g$ and $z_{2\varphi}^{-1}z_{1\varphi}g$ and involved in the covariant derivatives of the spinor and scalar fields coincide with the corresponding constant $\tilde{g} = z_2^{-1}z_1 g$, which figure in the Lagrangian of the Yang-Mills field:

$$z_{2\psi}^{-1}z_{1\psi} = z_{2\varphi}^{-1}z_{1\varphi} = z_2^{-1}z_1. \qquad (8.34)$$

As before, the conditions $z_{2\psi} = z_{1\psi}, z_{2\varphi} = z_{1\varphi}$ are not necessary. The gauge invariance does not impose any restrictions on the counterterms d and f, renormalizing the masses of the fields and the counterterm $z\lambda(\varphi^2)^2$. We shall choose the constants $z_{2\psi}$ and $z_{2\varphi}$ in accordance with the condition of finiteness of the two-point Green functions of the spinor and scalar fields.

If the condition (8.34) is fulfilled, then the Green functions generated by the Lagrangian (8.33) satisfy the generalized Ward identities (7.26) with the obvious substitution $g \to \tilde{g}$. The proof of the finiteness of the Green function repeats word for word the reasoning given above. The only difference is that the functional F on the right-hand side of (8.20) contains additional terms

$$\tilde{z}_1 g \int \zeta^b(y)(\Gamma^d)^{bc}\left[\frac{1}{i}\frac{\delta G^{da}_{yx}}{\delta\zeta^c(y)}\right] dy + \ldots, \qquad (8.35)$$

where ... denotes analogous terms for the spinor fields. The corresponding diagrams are presented in Fig. 23. The finiteness of these diagrams is demonstrated exactly in the same manner as the finiteness of the diagrams in Fig. 21.

All the remaining reasoning is entirely identical to the analogous reasoning for the Yang-Mills field in vacuum. Thus, in order to remove all the ultraviolet divergences, the gauge-invariant counterterms are sufficient in this case also.

Nor do any new features appear in the theory with spontaneously broken symmetry. The above-described scheme for proving the renormalizability remains unchanged. Consider, for example, the model (3.25) of Chapter 1. The most general form of an admissible renormalized Lagrangian can be obtained in the

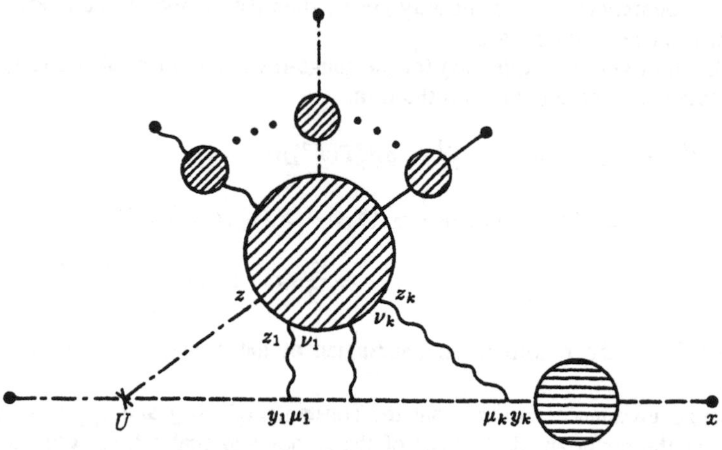

Figure 23.

following manner. In accordance with the procedure described above, admissible counterterms

$$\mathcal{L}' = z_{2\varphi}\left|\left(\partial_\mu\varphi^+ - iz_{1\varphi}z_{2\varphi}^{-1}g\frac{\tau^a}{2}A_\mu^a\varphi^+\right)\right|^2 - z\lambda^2(\varphi^+\varphi - \mu^2 + \delta\mu^2)^2 + \mathcal{L}_{YM}^R. \quad (8.36)$$

are inserted in the Lagrangian (3.25) of Chapter 1. The constants $z_{2\varphi}, z_{1\varphi}$ satisfy the conditions (8.34). Passing to the shifted fields B^a, σ defined by the formula (7.31), we obtain

$$\begin{aligned}\mathcal{L}_R = {} & \frac{z_{1\sigma}^2 z_{2\varphi}^{-1}}{2}m_1^2 A_\mu^2 + z_{1\varphi}m_1 A_\mu^a\partial_\mu B^a + \frac{1}{2}z_{2\varphi}\partial_\mu B^a\partial_\mu B^a \\ & + \frac{1}{2}z_{2\varphi}\partial_\mu\sigma\partial_\mu\sigma - \frac{m_2^2}{2}\sigma^2 - \frac{zm_2^2}{m_1^2}\frac{g^2}{8}\delta\mu^2(B^2 + \sigma^2) + \\ & - \frac{zm_2^2}{m_1}\frac{g}{2}\delta\mu^2\sigma + z_{1\varphi}\frac{g}{2}A_\mu^a(\sigma\partial_\mu B^a - B^a\partial_\mu\sigma - \varepsilon^{abc}B^b\partial_\mu B^c) \\ & + z_{1\varphi}^2 z_{2\varphi}^{-1}\left[\frac{m_1 g}{2}\sigma A_\mu^2 + \frac{g^2}{8}(\sigma^2 + B^2)A_\mu^2\right] \\ & - \frac{zgm_2^2}{4m_1}\sigma(\sigma^2 + B^2) - \frac{zg^2m_2^2}{32m_1^2}(\sigma^2 + B^2)^2. \quad (8.37)\end{aligned}$$

The renormalized Yang-Mills Lagrangian (8.13), involving also interaction with fictitious particles, remains unchanged, and we shall not write it out.

In passing to the formula (8.37) we performed a shift of the fields φ by a quantity equal to the vacuum mean of the field φ, not allowing for radiative corrections. Therefore in the Langrangian (8.37) there are present counterterms linear in the field σ, which compensate the divergences in the "tadpole"-type diagrams (Fig. 24), and also the counterterm renormalizing the masses of the Goldstone fields B^a. These counterterm renormalize the masses of the Goldstone fields B^a. These counterterms are necessary in order to provide for the equilibrium of the ground state when taking the radiative corrections into account.

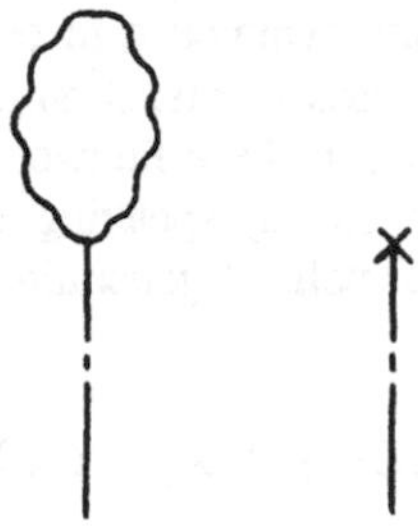

Figure 24. Diagrams of the "tadpole" type in the Yang-Mills theory with spontaneously broken symmetry.

The Lagrangian (8.36) is invariant with respect to the gauge transformations (7.32) with the substitution

$$m_1 \to \tilde{m}_1, \qquad g \to \tilde{g}, \qquad \tilde{g} = \tilde{z}_1 \tilde{z}\, g, \qquad \tilde{m}_1 = \tilde{z}_1 \tilde{z}_2^{-1} m_1 . \tag{8.38}$$

The generalized Ward identities are modified in the same way as in the symmetric case,

$$\begin{aligned}\frac{1}{\alpha} f^2(\Box)\partial_\mu \left[\frac{1}{i}\frac{\delta Z_R}{\delta J^a_\mu(x)}\right] &= \int \{D_0(x-y)\partial_\mu J^a_\mu(y)dy\} Z_R \\ &+ \int \Bigg\{ (J^b_\mu)^{\mathrm{tr}}(y)\tilde{z}_1 g \varepsilon^{bcd}\frac{1}{i}\frac{\delta}{\delta J^c_\mu(y)} + \frac{\tilde{z}_1 g}{2}\delta^{bd}\xi_\sigma(y)\frac{1}{i}\frac{\delta}{\delta \xi^b_B(y)} \\ &-\xi^b_B(y)\tilde{z}_1\left[\frac{g}{2}\varepsilon^{bcd}\frac{1}{i}\frac{\delta}{\delta\xi^c_B(y)} + \delta^{bd}\frac{g}{2}\frac{1}{i}\frac{\delta}{\delta\xi_\sigma(y)} + m_1\delta^{bd}\right]\Bigg\} \\ &\times G^{da}_R(J,\xi_B,\xi_\sigma,y,x)dy \end{aligned} \tag{8.39}$$

(we recall that in the gauges under consideration the constant $\tilde{z}_1$ is finite). The proof of the renormalizability repeats practically word for word the corresponding reasoning for the symmetric case. The only technical complication consists in that due to the mixing of the fields A^a_μ, B^a in the generalized gauge, the two-point Green functions are represented by matrices (2×2).

4.9 The Renormalized S-Matrix

We have shown that the renormalization procedure can be carried out without violating the gauge invariance of the theory. We shall now show that the renormalized theory obeys the relativity principle, meaning that the probabilities of physical processes do not depend on the actual choice of the gauge condition. Thus the unitarity of the renormalized S-matrix will be proved.

In the present section we shall consider models with spontaneously broken symmetry, in which all physical particles have non-zero masses. Formally, all the reasoning may be applied to the symmetric theory also; but in this case, as has already been pointed out, the matrix elements on the mass shell contain additional infrared singularities. Therefore, in the framework of perturbation theory, the S-matrix in the symmetric theory, strictly speaking, does not exist.

So let us consider the renormalized generating functional for the Green functions, which can be written as

$$Z(J_\mu, \zeta_\sigma) = N^{-1} \int \exp\left\{ i \int [\mathcal{L}_R + J_\mu^a A_\mu^a + \zeta_\sigma \sigma] dx \right\} \times \Delta(\mathcal{A}) \prod_x \delta(\partial_\mu \mathcal{A}_\mu) d\mathcal{A}\, d\sigma\, d\mathcal{B}. \quad (9.1)$$

Here $\mathcal{L}_R$ is the renormalized gauge-invariant Lagrangian of the Yang-Mills field interacting with the fields of matter. For definiteness we consider the Lagrangian (8.36). We shall assume the source J_μ^a to be transverse:

$$\partial_\mu J_\mu^a = 0. \quad (9.2)$$

The matrix elements of the S-matrix are expressed in terms of the variational derivatives Z by means of reduction formulas,

$$\begin{aligned} &S_{i_1 \ldots i_n, j_1 \ldots j_m}(k'_1 \ldots k'_n, p'_1 \ldots p'_l; k_1 \ldots k_m, p_1 \ldots p_q) V^{\frac{n+m}{2}} W^{\frac{l+q}{2}} \\ &\quad = (k_1'^2 - m_1^2) \ldots (k_n'^2 - m_1^2)(p_1'^2 - m_2^2) \ldots (p_l'^2 - m_2^2)(k_1^2 - m_1^2) \\ &\quad \ldots (k_m^2 - m_1^2)(p_1^2 - m_2^2) \ldots (p_q^2 - m_2^2)\theta(k'_{10}) \ldots \theta(k'_{n0})\theta(-k_{10}) \\ &\ldots \theta(-k_{m0})\theta(p'_{10}) \ldots \theta(p'_{l0})\theta(-p_{10}) \ldots \theta(-p_{q0}) u_{\mu_1}^{i_1} \ldots u_{\mu_n}^{i_n} G_{\mu_1 \ldots \mu_n \nu_1 \ldots \nu_m}(k'_1 \ldots p_1) \\ &\quad \times u_{\nu_1}^{j_1} \ldots u_{\nu_m}^{j_m} \Big|_{p^2 = p'^2 = m_2^2; k^2 = k'^2 = m_1^2}. \end{aligned} \quad (9.3)$$

Here k, k' denote the momenta of vector particles, and p, p' those of the scalar particles. The constants V and W are renormalizing factors:

$$\begin{aligned} &\delta^{ab}\left(g_{\mu\nu} - \frac{k_\mu k_\nu}{k^2}\right) \cdot V \underset{k^2 = m_1^2}{=} \\ &\quad = (k^2 - m_1^2)\left(g_{\mu\nu} - \frac{k_\mu k_\nu}{k^2}\right) \int e^{ikx} \frac{\delta^2 Z}{\delta J_\mu^a(x) \delta J_\nu^b(0)}\bigg|_{J,\zeta=0} dx, \end{aligned} \quad (9.4)$$

$$W \underset{p^2=m_2^2}{=} (p^2 - m_2^2) \int e^{ipx} \left. \frac{\delta^2 Z}{\delta\zeta_\sigma(x)\delta\zeta_\sigma(0)} \right|_{J,\zeta=0} dx. \tag{9.5}$$

If the two-point Green function is normalized on the mass shell to unity,

$$(p^2 - m^2)G(p^2) = 1,\ p^2 = m^2, \tag{9.6}$$

then these factors are absent, and we come back to the formula (3.64), Chapter 3. The matrix elements (9.3), calculated in the Lorentz gauge, tend to a definite limit when the intermediate regularization is removed. Let us demonstrate that in reality the values of the matrix elements (9.3) are independent of the choice of the gauge condition.

Let us pass in the expression (9.1) for the generating functional $Z(J, \zeta_\sigma)$ from the Lorentz gauge to the unitary gauge

$$\mathcal{B}^a = 0. \tag{9.7}$$

The invariance of the renormalized Lagrangian $\mathcal{L}_R$ allows us to use for this the same method as was used in Chapter 2.

By introducing the gauge-invariant functional $\Delta'(\sigma, \mathcal{B}, \tilde{g})$ defined by the equality

$$\Delta'(\sigma, \mathcal{B}, \tilde{g}) \int \delta(\mathcal{B}^\omega) \prod_x d\omega = 1, \tag{9.8}$$

where

$$\mathcal{B}^\omega = \mathcal{B} - \tilde{m}_1 u - \frac{\tilde{g}}{2}[\mathcal{B}, u] - \frac{\tilde{g}}{2}\sigma u + O(u^2), \tag{9.9}$$

we can rewrite the functional $Z(J, \zeta)$ in the form

$$Z(J_\mu, \zeta) = N^{-1} \int \exp\left\{ i \int [\mathcal{L}_R + J_\mu^a \mathcal{A}_\mu^a + \zeta_\sigma \sigma] dx \right\} \Delta(\tilde{g}, \mathcal{A})$$
$$\times \prod_x \delta(\partial_\mu \mathcal{A}_\mu) \Delta'(\mathcal{B}, \tilde{g}, \sigma) \delta(\mathcal{B}^\omega) d\omega\, d\mathcal{A}\, d\sigma\, d\mathcal{B}. \tag{9.10}$$

Passing to new variables

$$\mathcal{A}_\mu \to \mathcal{A}_\mu^{\omega^{-1}}, \quad \mathcal{B} \to \mathcal{B}^{\omega^{-1}}, \quad \sigma \to \sigma^{\omega^{-1}}, \quad \omega^{-1} \to \omega \tag{9.11}$$

and integrating over ω, we obtain, in complete analogy with the results of Chapter 2,

$$Z(J, \zeta) = N^{-1} \int \exp\left\{ i \int [\mathcal{L}_R + J_\mu^a (\mathcal{A}_\mu^\omega)^a + \zeta_\sigma \sigma^\omega] dx \right\}$$
$$\times \Delta'(\sigma, \mathcal{B}, \tilde{g}) \prod_x \delta(\mathcal{B}) d\mathcal{A}\, d\sigma\, d\mathcal{B}, \tag{9.12}$$

where

$$\begin{aligned} \mathcal{A}_\mu^\omega &= \mathcal{A}_\mu + \delta \mathcal{A}_\mu = \mathcal{A}_\mu + \partial_\mu u - \tilde{g}[\mathcal{A}_\mu, u] + O(u^2), \\ \sigma^\omega &= \sigma + \delta\sigma = \sigma - \frac{\tilde{g}}{2}(\mathcal{B}u) + O(u^2), \end{aligned} \tag{9.13}$$

and the function u is defined by the equation

$$\partial_\mu \mathcal{A}^\omega_\mu = \Box u - \bar{g}\partial_\mu[\mathcal{A}_\mu, u] + \partial_\mu \mathcal{A}_\mu + \ldots = 0. \tag{9.14}$$

The value of the functional $\Delta(\sigma, \mathcal{B}, \bar{g})$ on the surface $\mathcal{B} = 0$ is equal to

$$\Delta'(\sigma, \mathcal{B}, \bar{g}) \overset{=}{_{\mathcal{B}=0}} \det \left|\bar{m}_1 + \frac{\bar{g}\sigma(x)}{2}\right|^3 = \text{const} \det \left|m_1 + \frac{g\sigma(x)}{2}\right|^3. \tag{9.15}$$

The functional (9.12) differs from the generating functional for the Green functions in the unitary gauge only in the form of the terms with sources. We shall now show that if it is substituted in the reduction formula (9.3), then this difference vanishes; that means that the renormalized matrix elements remain unchanged for the substitution

$$J^a_\mu(\mathcal{A}^\omega_\mu)^a \to J^a_\mu \mathcal{A}^a_\mu, \quad \zeta_\sigma \sigma^\omega \to \zeta_\sigma \sigma. \tag{9.16}$$

The variational derivatives of the functional (9.12) are expressed in terms of Green functions of the form

$$\left(\frac{1}{i}\right)^{m+q} \frac{\delta^{(m+1)} Z}{\delta J^{a_1}_{\mu_1}(x_1) \ldots \delta J^{a_m}_{\mu_m}(x_m) \delta\zeta(y_1) \ldots \delta\zeta(y_q)}\bigg|_{j,\zeta=0}$$

$$= N^{-1} \int \exp\left\{i \int [\mathcal{L}_R] dx\right\} \Delta'(\sigma, \mathcal{B}, \bar{g})(\mathcal{A}^\omega_{\mu_1})^{a_1}(x_1)$$

$$\ldots (\mathcal{A}^\omega_{\mu_m})^{a_m}(x_m) \sigma^\omega(x_1) \ldots \sigma^\omega(x_q) \prod_x \delta(\mathcal{B}) d\mathcal{A}\, d\sigma\, d\mathcal{B}, \tag{9.17}$$

where $\mathcal{A}^\omega_\mu, \sigma^\omega$ are defined by the formulas (9.13). Since the sources $\quad_\mu$ are considered to be transverse, the linear term $\partial_{\mu\nu}$ does not give any contribution with the perturbation-theory expansion of $\delta\mathcal{A}_\mu$ and $\delta\sigma$ starts with terms quadratic in the fields.

Typical diagrams corresponding to Green functions are presented in Fig. 25. Diagrams of the types in (*a*) and (*b*) contain poles in all variables p_i, k_j. Diagrams of the type in (*c*) are one-particle irreducible at least in one of the momenta p_i, k_j. (The diagram presented in the figure is one-particle irreducible in the momentum p_1. This means that it is not possible to split it into two parts, connected only by one line, along which the momentum p_1 propagates.)

From the investigation of the analytical properties of Feynman diagrams, it is known that one-particle irreducible diagrams have no pole singularities in the corresponding variables. Therefore if the coefficient functions corresponding to the diagram in (*c*) are multiplied by the product

$$\prod_i (k_i^2 - m_1^2) \prod_j (p_j^2 - m_2^2) \tag{9.18}$$

and $k_1^2 = m_1^2, p_j^2 = m_2^2$ is assumed, then this expression vanishes. Diagrams of the (*b*) type are obtained from (*a*)-type diagrams by means of insertions in the external lines of the blocks as denoted in Fig. 25 at Π. On the mass shell this is equivalent

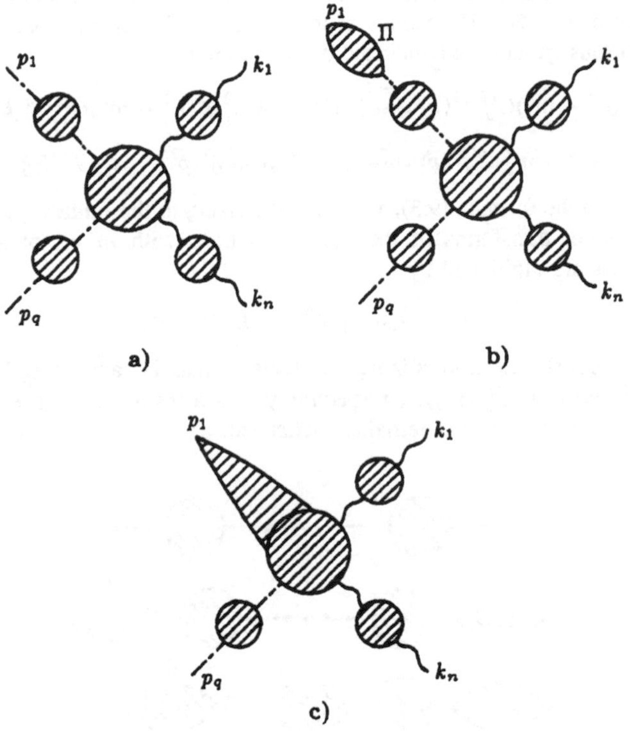

Figure 25.

to multiplying the corresponding Green functions by constants equal to the values of the functions $\Pi_A(k^2)$ and $\Pi_\sigma(p^2)$ at the points $k^2 = m_1^2, p^2 = m_2^2$. The values of the Green functions with m external vector and q external scalar lines on the mass shell when one gauge is changed to another are changed as follows:

$$\prod_{i=1}^{m}(k_i^2 - m_1^2)\prod_{j=1}^{q}(p_j^2 - m_2^2)G^{(u)a_1\ldots a_m}_{\mu_1\ldots\mu_m}(k_1,\ldots,k_m,p_1,\ldots,p_q)$$

$$= (1+\Pi_A(m_1^2)^n(1+\Pi_\sigma(m_2^2))^m \prod_{i=1}^{n}(k_i^2 - m_1^2)\prod_{j=1}^{q}(p_j^2 - m_2^2)$$

$$\times G^{(L)a_1\ldots a_m}_{\mu_1\ldots\mu_m}(k_1,\ldots,k_m,p_1,\ldots p_q)|_{\substack{k_i^2=m_1^2\\ p_j^2=m_2^2}}. \quad (9.19)$$

Here $G^{(u)}$ is the Green function in the unitary gauge, and $G^{(L)}$ is the Green function in the Lorentz gauge. Obviously, a quite analogous formula relates the Green functions in other gauges also.

The corresponding transformation for the two-point Green function is presented graphically in Fig. 26. The values of the two-point Green functions on the mass shell in various gauges are connected by the relation

$$(k^2 - m_1^2)G_{\mu\nu}^{(L)ab}(k) \underset{k^2=m_1^2}{=} (1 + \Pi_A(m_1^2))^2(k^2 - m_1^2)G_{\mu\nu}^{(u)ab}(k),$$

$$(p^2 - m_2^2)G_{\sigma}^{(L)}(p) \underset{p^2=m^2}{=} (1 + \Pi_{\sigma}(m_2^2))^2(p^2 - m_2^2)G_{\sigma}^{(u)}(p). \qquad (9.20)$$

Going back to the formula (9.3), we see that passing to the unitary gauge signifies that the Fourier transforms of the Green functions with m vector and q scalar external lines are multiplied by

$$(1 + \Pi_A(m_1^2))^m(1 + \Pi_{\sigma}(m_2^2))^q, \qquad (9.21)$$

but simultaneously the normalizing constants V and W are multiplied by $(1 + \prod_A(m_1^2))^2$ and $(1 + \prod_{\sigma}(m_2^2))^2$, respectively. As a result, the expression for the renormalized matrix element remains unchanged.

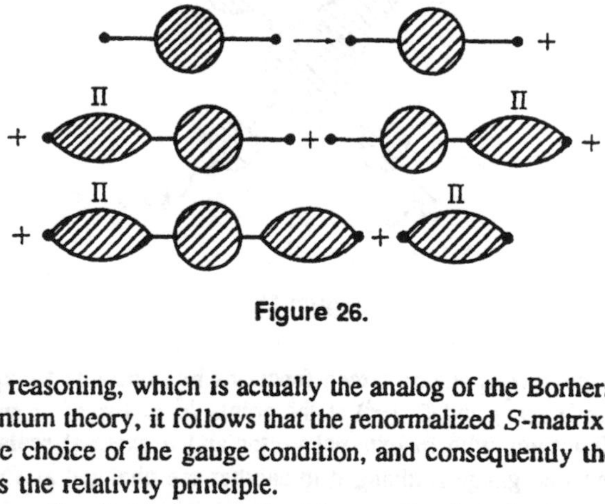

Figure 26.

From this reasoning, which is actually the analog of the Borhers theorm in the axiomatic quantum theory, it follows that the renormalized S-matrix is independent of the concrete choice of the gauge condition, and consequently the renormalized theory satisfies the relativity principle.

In the gauge $B = 0$ the renormalized Lagrangian has the form

$$\mathcal{L}_R = \mathcal{L}_{YM}^R + \left\{ \frac{z_{1\varphi}^2 z_{2\varphi}^{-1}}{2} m_1^2 A_\mu^2 + \frac{1}{2} z_{2\varphi}\partial_\mu\sigma\partial_\mu\sigma - \frac{(m_2^2 + \delta m_2^2)\sigma^2}{2} - \frac{2\delta m_2^2 m_1}{g}\sigma \right.$$
$$\left. + \frac{m_1 g}{2} z_{1\varphi}^2 z_{2\varphi}^{-1}\sigma A_\mu^2 + \frac{z_{1\varphi} z_{2\varphi}^{-1}}{8} g^2\sigma^2 A_\mu^2 - \frac{zgm_2^2}{4m_1}\sigma^3 - \frac{zg^2m_2^2}{32m_1^2}\sigma^4 \right\} \qquad (9.22)$$

All nonphysical particles (the Goldstone bosons, the fictitious c-particles, the longitudinal quanta of the vector field) are absent, and the unitarity of the scattering matrix is obvious. Due to the independence of the S-matrix of the gauge, the matrix elements on the mass shell tend to a definite limit when the regularization is removed. Note that this is not true of the Green functions of the mass shell, generated by the Lagrangian (9.22). The free Green function of the vector field corresponding to the Lagrangian (9.22) has the form

$$D_{\mu\nu} = \frac{-1}{(2\pi)^4} \frac{g_{\mu\nu} - k_\mu k_\nu m_1^{-2}}{k^2 - m_1^2} \tag{9.23}$$

and as $k \to \infty$ it tends to a constant. Calculating the divergence index of the diagram containing n_3 trident vector vertices, n_4 four-legged vertices, and L_{ex} external vector lines, we find

$$\omega = 4 + 4n_4 + 2n_3 - 2L_{ex}; \tag{9.24}$$

With the increase of n_i the number of types of divergent diagrams increases infinitely, that is, off the mass shell the theory is nonrenormalizable. Nevertheless, a finite number of the counterterms written out in the formula (9.22) are sufficent for the removal of the divergences from the matrix elements on the mass shell. Gauge-invariance leads to the physical equivalence between the explicitly renormalizable and explicitly unitary gauges, due to which the renormalized S-matrix has both these properties.

Obviously, these conclusions do not depend on the concrete model (9.22); they all apply equally to the model (3.13) of Chapter 1 and also to the models involving additional gauge-invariant interaction of fermions. Only the gauge invariance of the renormalized Lagrangian is essential.

To conclude this section we shall go back to the question of the proof of the finiteness of the Green functions in the generalized renormalizable gauge. Until now we have considered either the Lorentz gauge or a gauge in which the longitudinal part of the vector Green function decreases rapidly as $k \to \infty$. We shall now show that this condition is not necessary and that the counterterms of the form (8.3) provide for the finiteness of the Green functions in any renormalizable gauges, that is, in gauges for which the longitudinal part of the free Green function of the vector field decreases no more slowly than the transverse part, as $k \to \infty$. The simplest example of such a gauge is the gauge with $f(k^2) = 1$.

In any such gauge the divergent diagrams have a structure already discussed above: Only diagrams with one, two, three, and four external lines can diverge. As before we can choose the constants $Z_2, \tilde{z}_2, z_{2\varphi}, z_{2\psi}.\tilde{z}_1, z, \delta m$ so as to make all the two-point Green functions and the vertex functions $\Gamma_{\bar{c}cA}, \Gamma_{\sigma 4}$ finite, and determine the constants $z_1, z_{z\varphi}, z_{1\psi}$ by the invariance of the renormalized action:

$$z_1 z_2^{-1} = \tilde{z}_1 \tilde{z}_2^{-1} = z_{1\varphi} z_{2\varphi}^{-1} = z_{1\psi} z_{2\psi}^{-1}. \tag{9.25}$$

Let us show that ratios of the type

$$\frac{\Gamma_{A^3}}{(\Gamma_{AA})^{3/2}}, \qquad \frac{\Gamma_{\bar{\psi}\psi A}}{(\Gamma_{\bar{\psi}\psi})(\Gamma_{AA})^{1/2}} \text{ and so on,} \tag{9.26}$$

where all the external vector legs are considered to be transverse, on the mass shell do not depend on the gauge. Indeed, it follows from the formulas (9.19) and (9.20) that a transfer from one gauge to another changes the functions under consideration in the following manner (we omit the tensor structure):

$$\begin{aligned}\Gamma_{A^3}(k_1,k_2,k_3) &\to (1+\Pi_A(m_1^2))^3\Gamma_{A^3}(k_1,k_2,k_3),\\ \Gamma_{AA}(k) &\to (1+\Pi_A(m_1^2))^2\Gamma_{AA}(k), \quad k_i^2 = m_1^2.\end{aligned} \tag{9.27}$$

By substituting these expressions into the formula (9.26), we verify the invariance of this ratio. According to the above demonstration, in the Lorentz gauge all functions are finite. The function Γ_{AA} is finite due to the choice of constants z_i. Therefore in any gauge the function $\Gamma_{A^3}(k_1,k_2,k_3)$ is finite at $k_i^2 = m_1^2$. Since this function can diverge only logarithmically in renormalizable gauges, the finiteness of Γ_{A^3} at any k_i follows.

The finiteness of all the remaining Green functions is demonstrated absolutely analogously. We emphasize that now we are considering Green functions off the mass shell. On the mass shell, divergences are absent in all gauges, including the not explicitly renormalizable ones (that is to say, gauges for which the longitudinal part of the Green function of the vector field at large k behaves as $k^{2n}, n > -1$).

In the renormalizable gauge a finite number of invariant counterterms provide for the existence of the Green functions off the mass shell also. Then the concrete values of the counterterms depend, of course, on the choice of the gauge. Specifically, in the general gauge the constant $\tilde{z}_1$ is already not finite.

4.10 The S-Matrix in the Covariant Formalism

The scheme developed in the preceding sections for constructing the renormalized scattering matrix reduces to the following. Quantization is performed within a noncovariant (for instance, Coulomb) gauge in which the space of states includes only physical particles. Further, when making use of gauge invariance we pass to a covariant gauge, and in this gauge we perform renormalization. The compatibility of the renormalization procedure and the principles of relativity, proved above, permits us, now in the renormalized theory, to pass once again to the Coulomb gauge, thus demonstrating unitarity of the renormalized S-matrix.

The question arises whether it is not possible to avoid using noncovariant quantities at intermediate states and from the very beginning to work in the relativistically invariant gauge. In the case of quantum electrodynamics such a formalism is well known; this is the Gupta-Bleuler formalism. In this formalism the action in the Feynman gauge is chosen as the initial action:

$$S_{\text{G.B.}} = \int \left[-\frac{1}{4}\mathcal{F}_{\mu\nu}\mathcal{F}_{\mu\nu} + \frac{1}{2}(\partial_\mu A_\mu)^2 + \ldots\right] dy = \int \left[-\frac{1}{2}(\partial_\mu A_\nu)^2 + \ldots\right] dy. \tag{10.1}$$

Owing to the presence of the term fixing the gauge, the action (10.1) is not degenerate; and we can apply the standard quantization procedure to it, considering all the components A_μ to be independent. Canonical quantization leads to covariant commutation relations for the creation and annihilation operators,

$$[a_\mu(k)a_\nu^*(k')] = -g^{\mu\nu}\delta(k-k'). \tag{10.2}$$

Since the commutators of the creation and annihilation operators of time quanta a_0, a_0^*, differ in sign from the respective commutators for the spatial quanta a_i, the norm of the states containing an odd number of time quanta is negative. To avoid the appearance of negative probabilities in observable quantities, it is necessary to single out in this space a subspace of physical states possessing a positive norm. This is achieved by imposing on the admissible states the weak Lorentz condition

$$\partial_\mu \mathcal{A}_\mu^-(x)|\Phi\rangle = 0. \tag{10.3}$$

Here $\mathcal{A}_\mu^-(x)$, as usual, denotes the negative-frequency part of the operator $\mathcal{A}_\mu$.

The condition (10.3) guarantees fulfillment of the Lorentz condition for expectation values and provides for the absence of negative probabilities in observable quantities. It can be demonstrated that the expectation values of all observable quantities (energy, momentum, etc.) calculated with the aid the states (10.3) coincide with the respective values calculated with the aid of states involving only three-dimensional transverse physical photons.

For physical unitarity of the theory, it is necessary for the scattering matrix to transform states satisfying (10.3) into similar states. This occurs in electrodynamics. This is due to the longitudinal part of the electromagnetic field satisfying the free equation

$$\Box \partial_\mu \mathcal{A}_\mu = 0, \tag{10.4}$$

and therefore interaction does not lead to transitions between physical states and states with a negative norm.

As noted above, straightforward application of the Gupta-Bleuler formalism to the non-Abelian Yang-Mills field leads to a contradiction: the S-matrix is not unitary in the space of states satisfying the condition (10.3). The reason for this is that is the non-Abelian case the longitudinal part of the Yang-Mills field does not satisfy the free equation, and in the course of evolution the states (10.3) can transform into states with negative norm. It is necessary to find another relativistically invariant condition that on the one hand, would provide for the absence of negative probabilities and for observable quantities being identical with the respective quantities in the Coulomb gauge and, on the other hand, would be conserved in the course of evolution. Formulation of such a condition is rendered possible by invariance of the effective action with respect to the transformations (7.35) to (7.37) which are conventionally known as the BRS (Becchi-Rouet-Stora) transformations.

We shall choose as the initial action the effective action in the Feynman gauge (it is not difficult to consider also an arbitrary covariant α-gauge),

$$\mathcal{L} = -\frac{1}{4}F_{\mu\nu}^a F_{\mu\nu}^a - \frac{1}{2}(\partial_\mu A_\mu)^2 + i\bar{c}^a \partial_\mu(\nabla_\mu c)^a. \tag{10.5}$$

This Lagrangian reduces to the Lagrangian present in the formula (3.54) of Chapter 3 for the S-matrix by the change of variables $\bar{c} \to i\bar{c}$ for $\alpha = -1$. The Lagrangian $\mathcal{L}$ is explicitly Hermitian if the following rules for conjugating fictitious fields are adopted:

$$\bar{c}^+ = \bar{c}; \qquad c^+ = c. \tag{10.6}$$

Since the Lagrangian (10.5) is not degenerate, the standard procedure of canonical quantization can be applied to it. In this case all the variables $\mathcal{A}_\mu$ are independent; and, therefore, instead of the expansion (2.43) of Chapter 3, it is convenient to make use of the following expansion:

$$\mathcal{A}_\mu(x,t) = \frac{1}{(2\pi)^{3/2}} \int [e^{ikx} a^l(k,t) e^l_\mu + e^{-ikx} a^{*l}(k,t) e^l_\mu] \frac{d^3k}{\sqrt{2\omega}}, \tag{10.7}$$

where e^1 and e^2 are unit polarization vectors orthogonal to the momentum vector k and to each other: $e^3 = \frac{k}{|k|}$.

$$(e^l e^m) = \delta^{lm}, l,m = 1,2,3; \quad e^l_0 = 0; \quad e^0_\mu = \delta_{\mu 0}. \tag{10.8}$$

Under operator quantization the complex amplitudes a^* and a play the role of creation and annihilation operators, while the standard procedure of canonical quantization yields the following commutation relations:

$$[a_\mu(k,t), a^*_\nu(k',t)] = -g^{\mu\nu}\delta(k-k'). \tag{10.9}$$

As can be seen from this formula, the commutator of the operators of time quanta has a wrong sign, which leads to non-positive definiteness of the scalar product. The space of states in such theory is a space with indefinite metric.

Similar expansions for ghost fields with account of the hermiticity of (10.6) have the form

$$\begin{aligned} c(x,i) &= \frac{1}{(2\pi)^{3/2}} \int [e^{ikx} d(k,t) + e^{-ikx} d^*(k,t)] \frac{d^3k}{\sqrt{2\omega}}, \\ \bar{c}(x,t) &= \frac{1}{(2\pi)^{3/2}} \int [e^{ikx} \bar{d}(k,t) + e^{-ikx} \bar{d}^*(k,t)] \frac{d^3k}{\sqrt{2\omega}}. \end{aligned} \tag{10.10}$$

Canonical quantization leads to the following commutation relations:

$$\begin{aligned} [d(k,t), \bar{d}^*(k',t)]_+ &= \delta(k-k'), \\ [\bar{d}(k,t), d^*(k',t)]_+ &= \delta(k-k'). \end{aligned} \tag{10.11}$$

The operators for asymptotic fields are obtained from (10.7) and (10.10) by the usual change of variables:

$$\begin{aligned} a^l(k,t) &\to a^l(k)e^{-i\omega t}, & a^{*l}(k,t) &\to a^{*l}(k)e^{i\omega t}, \\ d(k,t) &\to d(k)e^{-i\omega t}, & d^*(k,t) &\to d^*(k)e^{i\omega t}, \\ \bar{d}(k,t) &\to \bar{d}(k)e^{-i\omega t}, & \bar{d}^*(k,t) &\to \bar{d}^*(k)e^{i\omega t}. \end{aligned} \tag{10.12}$$

The corresponding asymptotic space is, as already noted above, a space with indefinite metric; and it realizes a representation of the algebra of BRS transformations.

From invariance of the Lagrangian (10.5) with respect to the transformations (7.35) to (7.37), in which the change of variables $\bar{c} \to i\bar{c}$ must be performed, follows the conservation of current:

$$\mathcal{Y}_\mu^B = F_{\mu\nu}^a(\nabla_\nu c)^a - (\partial_\nu A_\nu^a)(\nabla_\nu c)^a - \frac{i}{2}\partial_\mu \bar{c}^a t^{abd} c^b c^d. \tag{10.13}$$

The charge

$$Q_B = \int \mathcal{Y}_0^B d^3x \tag{10.14}$$

is conserved in time and is a generator of the BRS transformations (7.35) to (7.37).

The Lagrangian (10.5) generates one more conservation law, which follows from the invariance of (10.5) with respect to the scale transformations

$$c \rightarrow e^\alpha c, \qquad \bar{c} \rightarrow e^{-\alpha}\bar{c}. \tag{10.15}$$

The corresponding conserved current $\mathcal{Y}_\mu^c$ has the form

$$\mathcal{Y}_\mu^c = i(\bar{c}^a \nabla_\mu c^a - \partial_\mu \bar{c}^a c^a). \tag{10.16}$$

The charge

$$Q_c = \int \mathcal{Y}_0^c d^3x \tag{10.17}$$

is called a ghost charge. The eigen-values of the operator iQ_c are called ghost numbers. Clearly,

$$\begin{aligned} [iQ_c, c(x)] &= c(x), \\ [iQ_c, \bar{c}(x)] &= -\bar{c}(x). \end{aligned} \tag{10.18}$$

The operators Q_B and Q_c realize a BRS algebra,

$$[Q_B, Q_B]_+ = 2Q_B^2 = 0, \tag{10.19}$$

$$[iQ_c, Q_B] = Q_B, \tag{10.20}$$

$$[Q_c, Q_c] = 0. \tag{10.21}$$

From the formula (10.19) it can be seen that the squared operator Q_B equals zeero. Such operators are called nilpotent. The equality (10.19) is readily verified directly:

$$\begin{aligned} 2i\varepsilon Q_B^2 &= \left[i\varepsilon Q_B, \int d^3x \int F_{0\nu}^a(\nabla_\nu c)^a - \partial_\nu A_\nu^a(\nabla_0 c)^a - \frac{i}{2}\partial_0 \bar{c}^a t^{abd} c^b c^d\right] \\ &= \int d^3x \left[t^{abd} F_{0\nu}^b c^d (\nabla_\nu c)^a - \frac{1}{2}\partial_0(\partial_\nu A_\nu)^a t^{abd} c^b c^d\right] = 0 \end{aligned} \tag{10.22}$$

Here we have taken advantage of the fact that $\delta(\nabla_\mu c) = 0$ and $\delta(t^{abd}c^b c^d) = 0$ (see (7.39)), as well as of the equations of motion.

The equalities (10.20) and (10.21) signify that the charges Q_B and Q_c exhibit the ghost numbers 1 and 0, respectively.

Let us single out the physical sector by the conditions

$$Q_B|\Phi\rangle = 0, \tag{10.23}$$

$$Q_c|\Phi\rangle = 0. \tag{10.24}$$

Since the operators Q_B and Q_c commute with the Hamiltonian, the subspace defined by the equations (10.23) and (10.24) is invariant with respect to time translations, and the scattering matrix transforms physical asymptotic states into physical ones. For asymptotic states the conditions (10.23) and (10.24) are simplified:

$$\lim_{|t|\to\infty} e^{iH_0t}Q_{B(c)}e^{-iH_0t} = Q^0_{B(c)}, \tag{10.25}$$

where

$$Q^0_B = \int d^3x[(\partial_\nu A_0 - \partial_0 A_\nu)^a\partial_\nu c^a - (\partial_\nu A^a_\nu)\partial_0 c^a], \tag{10.26}$$

$$Q^0_c = \int d^3x i[\bar{c}^a\partial_0 c^a - \partial_0\bar{c}^a c^a]. \tag{10.27}$$

We have to construct asymptotic states that are annihilated by the operators Q^0_B and Q^0_c. To this end we express the operator Q^0_B in terms of creation and annihilation operators. Substituting the expansions (10.7) and (10.10) into (10.26), we obtain

$$Q^0_B = -\int d^3k\omega[(a_0^* + a_3^*)d + d^*(a_0 + a_3)]. \tag{10.28}$$

The condition (10.24) signifies that in a physical state the number of ghosts (d) equals the number of antighosts ($\bar{d}$).

Since the operator Q^0_B is independent of the transverse components of the field, any state $|\Phi_{tr}\rangle$ containing only transversely polarized quanta satisfies the condition (10.23), and an arbitrary state satisfying (10.23) and (10.24) can be represented as the sum of direct products $|\Phi_{tr}\rangle \bigotimes |\tilde{\Phi}\rangle$ where $|\tilde{\Phi}\rangle$ has the form

$$\tilde{\Phi}\rangle = \sum_{n,k,m} \int dp_j C^{n,k}_m(p_j) \prod_{j_1=0}^{n} a^*_+(p_{j_1}) \prod_{j_2=0}^{k} a^*_-(p_{j_2}) \times \prod_{j_3,j_4=0}^{m} d^*(p_{j_3})\bar{d}^*(p_{j_4})|0\rangle. \tag{10.29}$$

Here $C^{n,k}_m(p_j)$ are coefficient functions depending on the arguments $p_{j_i}(i = 1,\ldots,4)$; $a^*_+(p)$ and $a^*_-(p)$ denote the combinations

$$a^*_+(p) = \frac{a_0^*(p) + a_3^*(p)}{\sqrt{2}}; \qquad a^*_-(p) = \frac{a_0^*(p) - a_3^*(p)}{\sqrt{2}}. \tag{10.30}$$

The condition for annihilation of the vector $|\tilde{\Phi}\rangle$ by the operator Q^0_B has the form

$$\begin{aligned} &(m+1)C^{n,k}_{m+1} = -(k+1)C^{n+1,k+1}_m; \\ &C^{0,k+1}_m = 0; \qquad m,n,k = 0\ldots\infty. \end{aligned} \tag{10.31}$$

Taking into account that

$$\begin{aligned} &[a^*_+, a_+] = [a^*_-, a_-] = 0, \\ &[a_+(k), a^*_-(k')] = -\delta(k - k'), \end{aligned} \tag{10.32}$$

we find that the norm of the vector $|\bar{\Phi}\rangle$ is

$$\langle\bar{\Phi}|\bar{\Phi}\rangle = \sum_{n,k,m}(-1)^{m+n+k}\int C_m^{*n,k} C_m^{k,n} dp_j n!k!(m!)^2 \tag{10.33}$$

Owing to the condition (10.31), all the terms in this expression, except the vacuum term corresponding to $n = k = m = 0$, compensate each other pairwise. Consequently, any vector satisfying the annihilation condition (10.23) has the form $|\Phi^{tr}\rangle \bigotimes |\Phi_0\rangle$ where $|\Phi_0\rangle$ is a vector with a zero norm (null-vector). From the Cauchy-Bunyakovsky inequality it follows that the vectors $|\Phi_0\rangle$ are othogonal to all other vectors. They, therefore, yield no contribution to the matrix elements, and any vectors differing by $|\Phi_0\rangle$ can be considered identical.

Thus, the norm of a physical state $\langle\Phi|\Phi\rangle$ equals $\langle\Phi_{tr}|\Phi_{tr}\rangle$. In spite of the presence in the total space of states of negative norm, all the physical vectors have a positive norm determined only by the three-dimensionally transverse components. According to the above proof the S-matrix is unitary in the physical sector.

Clearly, the expectation value of any observable quantity in the physical states is also determined only by the three-dimensionally transverse sector and coincides with the corresponding value in the Coulomb gauge. This follows from observables being gauge-invariant, and, consequently, the respective operators commute with the generator of the BRS transformation. The construction described above was developed for the Yang-Mills theory, and we assumed the perturbation theory to be applicable. However, a similar construction may turn out to be useful, also, for investigating the possibility of going beyond the scope of perturbation theory, as well as for describing other gauge-invariant systems, such as, for example, relativistic strongs. Therefore, to conclude this section, we shall formulate certain general results concerning the representations of the BRS algebra, (10.19) to (10.21).

Owing to the operator Q^B being nilpotent, the representations of the BRS algebra have a maximum dimensionality equal to two. There exist irreducible singlet and doublet representations. A singlet state is a state satisfying the condition $Q_B|s\rangle = 0$ that cannot be represented in the form $|s\rangle = Q_B|f\rangle$. A doublet is a pair of states $|p\rangle$ and $|d\rangle$ satisfying the conditions $|d\rangle = Q_B|p\rangle, Q_B|d\rangle = 0$. The states $|p\rangle$ and $|d\rangle$ are called parent and daughter states, respectively.

This classification is not quite unambiguous since, for example, if $|s\rangle$ is a singlet state, then the state $|s\rangle + |d\rangle$ will also be a singlet. The ambiguity can be removed by fixing the basis in the space in which the representation of the BRS algebra acts. There exists a basis in which the entire space V is decomposed into a direct sum of three subspaces:

$$V = V_p \bigoplus V_d \bigoplus V_s. \tag{10.34}$$

The singlet subspace V_s is singled out by the condition

$$qV_s = q^+ V_s = \{0\}. \tag{10.35}$$

Here q is a matrix representing Q_B in the chosen basis. The condition of Hermitian conjugation for q has the form $\eta q = q^+ \eta$, where η is the metric determined in the usual way: $\eta_{kl} = \langle e_k | e_l \rangle$, $|e_k\rangle$ are base vectors.

The daughter and parent subspaces are singled out by the conditions

$$qV_d = q^+ V_p = 0, \tag{10.36}$$

$$qV_p = V_d; \qquad q^+ V_d = V_p. \tag{10.37}$$

The space in which the representation of the BRS algebra acts is a space with indefinite metric. The physical space is singled out by the condition $Q_B|\Phi\rangle = 0$ and permits explicit representation through the above introduced subspaces,

$$V_{ph} = V_s \bigoplus V_d. \tag{10.38}$$

Clearly, in the case of arbitrary physical states the scalar product depends only on the singlet components,

$$\langle \Phi | \tilde{\Phi} \rangle = \langle \Phi | P(V_s) | \tilde{\Phi} \rangle, \tag{10.39}$$

where $P(V_s)$ is the projector onto the singlet subspace. Hence it follows that physical states have a positive norm only in that case when the singlet subspace has a positive definite metric. The latter is determined by the concrete model chosen and requires special investigation. If the theory is invariant with respect to the BRS transformations, then Q_B commutes with the Hamiltonian, and the S-matrix is unitary in the physical space. The matrix elements, then, depend, in practice, only on the singlet components. However, we shall once again stress that BRS invariance alone is still insufficient for the theory to be physically admissible. In the general case "physical" vectors that are annihilated by the generator of BRS transformations may have a negative norm. Therefore an additional investigation of the BRS-singlet sector of the model is necessary.

4.11 Anomalous Ward Identities

In the construction of the unitary renormalized S-matrix we use the invariant intermediate regularization. The existence of the invariant regularized action allowed us to obtain the generalized Ward identities and, using them, to prove the physical equivalence of the unitary and Lorentz gauges. Generally speaking, it is not necessary to employ the invariant intermediate regularization and try to choose the counterterms in such a manner that the renormalized Green functions would satisfy the generalized Ward identities. For this, if the regularization is noninvariant, noninvariant counterterms such as the photon mass renormalization in electrodynamics may be required.

In this case, in the regularized theory the relativity principle is violated, and its correctness in the limit when the intermediate regularization is removed requires a special investigation. It may turn out that whatever the choice of local counterterms, the renormalized Green functions will not satisfy the generalized Ward identities. This will lead to the nonequivalence of different gauges and the inconsistency of the theory. In this case the unitary renormalized S-matrix does not (at least in the framework of perturbation theory) exist.

In practice the indicated situation arises when the matrix γ_5 is involved in gauge transformations of the fermion fields. In this case both the above described methods for invariant regularization are inapplicable. In the framework of the dimensional regularization it turns out not to be possible to give a consistent definition of the matrix γ_5 for a space with arbitrary dimensionality. The regularization by means of the higher covariant derivatives still provides the finiteness of all multiloop diagrams; however, the invariant regularization of one-loop diagrams by means of the Pauli-Villars procedure in this case is impossible, since the mass terms for the fermion fields $\mu_j \bar{\psi}_j \psi_j$ violate the γ_5-invariance. Thus, for one-loop diagrams there is no γ_5-invariant regularization, and, as we see, for a number of gauge groups involving γ_5-transformations, the Green functions do not satisfy the generalized Ward identities.

As a simple, example, consider a model with the $U(1)$ gauge group, described by the Lagrangian

$$\mathcal{L} = -\frac{1}{4}(\partial_\nu A_\mu - \partial_\mu A_\nu)^2 + i\bar{\psi}\gamma_\mu(\partial_\mu - igA_\mu\gamma_5)\psi, \tag{11.1}$$

$$\gamma_5 = -i\gamma_0\gamma_1\gamma_2\gamma_3.$$

This Lagrangian is invariant under the Abelian gauge transformations

$$\begin{aligned} A_\mu(x) &\to A_\mu(x) + \partial_\mu\lambda(x), \\ \psi(x) &\to e^{ig\gamma_5\lambda(x)}\psi(x); \\ \bar{\psi}(x) &\to \bar{\psi}(x)e^{ig\gamma_5\lambda(x)}, \end{aligned} \tag{11.2}$$

and at first sight all the reasoning for the equivalence of various gauges can be equally applied to it. In the α-gauge the effective Lagrangian has the form

$$\mathcal{L}_{\text{ef}} = \mathcal{L} + \frac{1}{2\alpha}(\partial_\mu A_\mu)^2, \tag{11.3}$$

where $\mathcal{L}$ is the gauge-invariant expression (11.1). The Lagrangian (11.3) is nondegenerate and describes not only transversely polarized quanta of the vector field but also scalar quanta of spin zero.

One might take the Lagrangian (11.3) as a starting point and base the construction of a quantum theory on it. It is well known that such a theory would be inconsistent physically: The probability of events involving scalar quanta can take negative values. If, however, the Green functions generated by the Lagrangian (11.3) satisfy the Ward identities

$$\left\{\frac{1}{\alpha}\partial_\mu\left(\frac{\delta Z}{\delta J_\mu(x)}\right) - Z\partial_\mu J_\mu(x) + ig\bar{\eta}(x)\gamma_5\frac{\delta Z}{\delta\bar{\eta}(x)} + ig\frac{\delta Z}{\delta\eta(x)}\gamma_5\eta(x)\right\} = 0, \tag{11.4}$$

where Z is a generating functional of the form

$$Z = N^{-1} \int \exp\left\{ i \int [\mathcal{L}_{ef} + J_\mu A_\mu + \bar{\eta}\psi + \bar{\psi}\eta]\, dx \right\} dA d\bar{\psi} d\psi, \qquad (11.5)$$

then, as it is easy to show, the matrix elements of transitions between states involving transversely polarized quanta and states involving scalar quanta is equal to zero. This means that the S-matrix connecting "physical" transversely polarized states is unitary. (Strictly speaking, the S-matrix does not exist in the model considered, due to infrared divergences. It can be shown, however, that all the reasoning may be applied to the case when the vector field has a nonzero mass and the infrared divergences are absent.)

Formally, the identity (11.9) follows from the invariance of the Lagrangian (11.9) under the transformations (11.9). A specific case is the relation

$$\partial_{\mu_1}^{x_1} \frac{\delta^n \ln Z}{\delta J_{\mu_1}(x_1) \ldots \delta J_{\mu_n}(x_n)}\bigg|_{J,\eta=0} = 0, \qquad n > 2, \qquad (11.6)$$

which demonstrates explicitly the absence of transitions between transversely and longitudinally polarized states. In reality we are interested not in the naive identities (11.4), which strictly speaking, have no sense because of the divergent integrals involved in them, but in the corresponding relations for the renormalized Green functions. In electrodynamics, as also in the non-Abelian models discussed above, the Green functions satisfy generalized Ward identities that differ from the "naive" ones only by the renormalization of the charges and masses involved. This is not so in the model (11.1). The Green function with three external vector lines, corresponding to the diagram presented in Fig. 28, does not satisfy the "naive" identities (11.4), no matter what local counterterms have been chosen. The identity (11.6) means that the Fourier transform of the three-point vertex function $\Gamma_{\mu\nu\alpha}(p,q)$, defined by the equality

$$\Gamma_{\mu\nu\alpha}(p,q) G_{\mu\mu'}(p) G_{\nu\nu'}(q) G_{\alpha\alpha'}(p+q)$$
$$= \int e^{ipx} e^{iqy} \left(\frac{-i\delta^3 Z}{\delta J_{\mu'}(x) \delta J_{\nu'}(y) \delta J_{\alpha'}(0)} \right) dx\, dy, \qquad (11.7)$$

must be transverse:

$$p_\mu \Gamma_{\mu\nu\alpha}(p,q) = q_\nu \Gamma_{\mu\nu\alpha}(p,q) = (p+q)_\alpha \Gamma_{\mu\nu\alpha}(p,q) = 0. \qquad (11.8)$$

The explicit calculation of $\Gamma_{\mu\nu\alpha}(p,q)$, taking into account that the function $\Gamma_{\mu\nu\alpha}(p,q)$ is symmetric in the arguments $(p,\mu), (q,\nu), (-(p+q),\alpha)$, gives

$$i(p+q)_\alpha \Gamma_{\mu\nu\alpha}(p,q) = -\frac{g^3}{6\pi^2} \varepsilon^{\mu\nu\alpha\beta} p_\alpha q_\beta \neq 0. \qquad (11.9)$$

Since the index of the diagram in Fig. 27 is equal to unity, the function $\Gamma_{\mu\nu\alpha}(p,q)$ is defined to the approximation of a first-order polynomial in p and q. One might try to use this arbitrariness in order to set the right-hand side of the equality (11.9) equal to zero. It is easy to see, however, that this is impossible. The most general expression for a renormalized three-legged vertex function has the form

$$\tilde{\Gamma}_{\mu\nu\alpha}(p,q) = \Gamma_{\mu\nu\alpha}(p,q) + c_1\varepsilon_{\mu\nu\alpha\beta}p_\beta + c_2\varepsilon_{\mu\nu\alpha\beta}q_\beta, \tag{11.10}$$

where $\Gamma_{\mu\nu\alpha}(p,q)$ is a symmetric vertex function satisfying the relation (11.9). By requiring the function $\tilde{\Gamma}_{\mu\nu\alpha}$ also to be symmetric in the arguments (μ,p), (ν,q), $(\alpha,-(p+q))$, we obtain

$$c_1 = c_2 = 0. \tag{11.11}$$

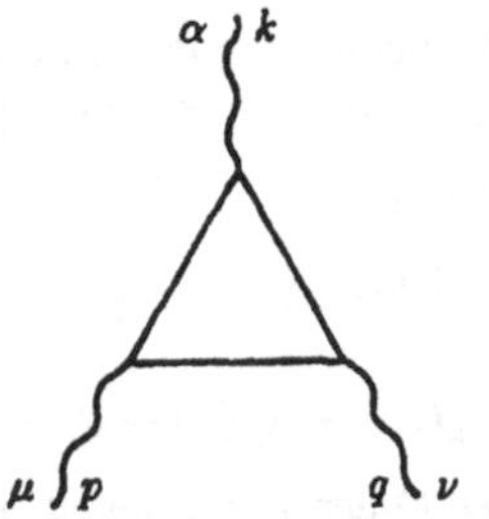

Figure 27. Anomalous triangle diagram.

Thus, there is no possible choice of the local counterterms to make the renormalized vertex function satisfy the identity (11.8). In consequence, the probability of transition from transversely polarized states to longitudinal ones is not equal to zero. The model described by the Lagrangian (11.1) is inconsistent. The stated difficulty is inherent in all the theories invariant under the Abelian gauge transformations involving the matrix γ_5. However, there exists a class of models for which this difficulty can be avoided. For instance, in the model (11.1), suppose that besides the field ψ, a field ψ' is involved, which interacts with a vector field in the same way as ψ but differs from the latter in the sign of the charge:

$$\mathcal{L} = -\frac{1}{4}(\partial_\nu A_\mu - \partial_\mu A_\nu)^2 + i\bar{\psi}\gamma_\mu(\partial - igA\gamma_5)\psi + i\bar{\psi}'\gamma_\mu(\partial + igA\gamma_5)\psi'. \tag{11.12}$$

Then together with the diagram in Fig. 27 there is an analogous diagram, in which the internal lines correspond to the fields ψ'. From the formula (11.9) it may be seen that the divergence of the anomalous vertex function is proportional to g^3. Therefore, the diagram corresponding to ψ' will give the same contribution to the identity (11.9), but with opposite sign. As a result, the total vertex function $\Gamma_{\mu\nu\alpha}$ will satisfy the normal identities (11.8). With regard to the diagrams containing more than one loop, the absence of anomalies can be proved for them in a general form. Indeed, as was shown in Section 4.3, regularization by means of the higher covariant derivatives makes all multiloop diagrams in any arbitrary gauge-invariant theory convergent. Therefore, if there are no anomalies in the one-loop diagrams, then the multiloop diagrams without doubt satisfy the normal Ward identities. The absence of anomalies in the model (11.12) can be explained also in the following way. It is possible to pass, in the Lagrangian (11.12), to new canonical variables

$$\psi_1 = \frac{1}{2}\{(1-\gamma_5)\psi + (1+\gamma_5)\psi'\},$$
$$\psi_2 = \frac{1}{2}\{(1+\gamma_5)\psi + (1-\gamma_5)\psi'\}. \tag{11.13}$$

The interaction Lagrangian expressed in terms of the fields ψ_1, ψ_2 does not contain the matrices γ_5,

$$\mathcal{L}_1 = (g\bar{\psi}_1\gamma_\mu\psi_1 - g\bar{\psi}_2\gamma_\mu\psi_2)A_\mu, \tag{11.14}$$

and represents an analog of the electromagnetic interaction Lagrangian of two massless spinors. Such a theory is invariant under purely vector gauge transformations

$$\psi_1(x) \to e^{ig\alpha(x)}\psi_1(x), \qquad \bar{\psi}_1(x) \to e^{-ig\alpha(x)}\bar{\psi}_1(x),$$
$$\psi_2(x) \to e^{-ig\alpha(x)}\psi_2(x), \qquad \bar{\psi}_2(x) \to e^{ig\alpha(x)}\bar{\psi}_2(x), \tag{11.15}$$
$$A_\mu(x) \to A_\mu(x) + \partial_\mu\alpha(x),$$

and therefore the renormalized Green functions satisfy the normal Ward identities. Such a mechanism for compensating anomalies may be used also in more realistic models, specifically, in models with spontaneously broken symmetry. If the fields ψ, ψ', A_μ interact also with scalar fields, then with a corresponding choice of potential all physical particles may be made to acquire zero masses due to the Higgs mechanism. At the same time the form of interaction of the spinor and vector fields responsible for the appearance of anomalies remains unchangled. Therefore, all the reasoning concerning the compensation of anomalies remains correct.

The appearance of anomalies in the Ward identities can be interpreted as a violation of the corresponding classical symmetry. The anomalous identity (9.9) can be written in the form

$$\int \exp\left\{i\int \mathcal{L}(x)dx\right\}\left\{\partial_\mu j^A_\mu(x) + \frac{g^2}{48\pi^2}\varepsilon^{\mu\nu\rho\sigma}F_{\mu\nu}F_{\rho\sigma}(x)\right\} \times d\bar{\psi}d\psi dA = 0, \tag{11.16}$$

where $\mathcal{L}$ is determined by the formula (11.1), $F_{\mu\nu}$ is the usual field strength tensor for the field A_μ, and j^A_μ is the axial current: $j^A_\mu = \bar{\psi}\gamma_\mu\gamma_5\psi$. In classical theory the axial current j^A_μ with which the field A_μ interacts is conserved. From the equality (11.16) it follows that in quantum theory the divergence $j^A_\mu \neq 0$. Nevertheless in this case, also, it is possible to define the conserved current $\tilde{j}^A_\mu$:

$$\tilde{j}^A_\mu = j^A_\mu + \frac{g^2}{12\pi^2}\varepsilon^{\mu\nu\rho\sigma}A_\nu\partial_\rho A_\sigma. \tag{11.17}$$

This current, however, does not coincide with the current with which the field A_μ interacts, and its conservation does not provide for invariance of the theory with respect to the gauge transformations (11.2).

A natural question arises whether it is possible to modify the Lagrangian (11.1) by adding to it a term depending on the fields A_μ to provide for its variation to compensate exactly the anomaly and for the action as a whole to be gauge invariant. It is not difficult to verify in a straightforward way that this is not possible. The anomaly cannot be represented in the form of the variation of a local functional depending only on the fields A_μ.

There exist arguments in favor of changing the gauge field quantization procedure itself in the presence of anomalies since certain classical constraints of the first class may become constraints of the second class in anomalous quantum theory. This in turn may lead to the effective arising of novel non-classical degrees of freedom. Attempts to construct a consistent quantization procedure taking into account such arguments have not yet proven successful. In the example considered above, the appearance of anomalies leads to various gauges being non-equivalent and, consequently, to inconsistency of the theory. Anomalies, however, also may be present in consistent theories, for example, in quantum electrodynamics. In electrodynamics we can also define the axial current $j_\mu^A = \bar{\psi}\gamma_\mu\gamma_5\psi$, and in the limit of zero electron mass this current is conserved in classical theory.

The transition matrix element of the axial current into two photons in the lowest order of perturbation theory is described by a diagram (Fig. 28) differing from the corresponding diagram in the model (11.1) only by the absence of γ_5-matrices at the vertices A and B. It easy to understand that this difference is insignificant since by commuting the γ-matrices we can transpose both γ_5-matrices to a single vertex, upon which the matrix elements will become identical. Therefore, in electrodynamics the Ward identity for the divergence of the axial current in the limit of zero electron mass assumes a form similar to (11.9):

$$i(p+q)_\alpha \Gamma_{\mu\nu\alpha} = \frac{e^2}{8\pi^2}\varepsilon_{\mu\nu\alpha\beta}p_\alpha q_\beta. \tag{11.18}$$

The only difference consists in that in this case no symmetry exists with respect to the transposition of the vertices, $A \leftrightarrow C$ and $B \leftrightarrow C$. Therefore one may attempt to redefine the vertex function taking advantage of the counterterm arbitrariness in accordance with (11.10). Choosing $c_1 = -c_2 = -\frac{ie^2}{16\pi^2}$ we provide for the following condition to be fulfilled:

$$(p+q)_\alpha \tilde{\Gamma}_{\mu\nu\alpha} = 0. \tag{11.19}$$

The function $\tilde{\Gamma}_{\mu\nu\alpha}$, however, does not satisfy the condition

$$\partial_\mu \tilde{\Gamma}_{\mu\nu\alpha} = \partial_\nu \tilde{\Gamma}_{\mu\nu\alpha} = 0, \tag{11.20}$$

which represents a consequence of the conservation of the electromagnetic current $\bar{\psi}\gamma^\mu\psi$.

Thus, in electrodynamics it is not possible to provide simultaneously for the conservation of the electromagnetic and axial currents. Since conservation of the electromagnetic current is a well established experimental fact, we are compelled to draw the conclusion that in quantum electrodynamics the axial current is not conserved.

In this case axial anomaly does not lead to the theory being inconsistent since gauge invariance in electrodynamics is related to conservation of the vector elec-

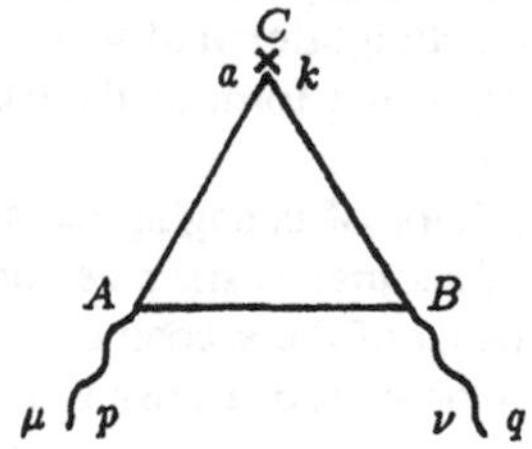

Figure 28. Anomalous triangle diagram in electrodynamics. The cross indicates the vertex corresponding to the axial current.

tromagnetic current, while from the point of view of this model the axial current represents an external object.

In electrodynamics with massive electrons, the axial anomaly leads to modification of the "partial" conservation law for the axial current. Instead of the classical equation

$$\partial_\mu j^A_\mu = 2mj_5, \tag{11.21}$$

where $j_5 = i\bar{\psi}\gamma_5\psi$, we obtain

$$\int \exp\left\{i\int \mathcal{L}(x)dx\right\}\left\{\partial_\mu j^A_\mu(x) - 2mj_5 + \frac{e^2}{16\pi^2}\varepsilon^{\mu\nu\rho\sigma}F_{\mu\nu}F_{\rho\sigma}\right\}d\psi dA = 0. \tag{11.22}$$

The anomalous identity (11.22) gives rise to important experimental consequences. Thus, for instance, the electromagnetic decay of the neutral pion into two photons is described by the anomalous diagram depicted in Fig. 28. Calculations performed on the basis of the identity (11.22) are in total agreement with experimental results.

Anomalous Ward identities may also appear in non-Abelian gauge fields. For example, let the spinor fields ψ interact in a gauge-invariant manner with the Yang-Mills field:

$$\mathcal{L} = i\bar{\psi}\gamma_\mu(\partial_\mu - g\Gamma^a A^a_\mu)\psi + \ldots \tag{11.23}$$

and let $\ldots$ stand for the Lagrangian of the Yang-Mills field and also, possibly, for the gauge-invariant interaction of the fields A_μ, ψ with scalar fields. The latter can correspond both to the symmetric theory and to the theory with spontaneously broken symmetry.

The matrices Γ^α realize the representation of the Lie algebra

$$[\Gamma^a, \Gamma^b] = t^{abc}\Gamma^c \tag{11.24}$$

and can also include the matrix γ_5. The divergence of the three-legged vertex Green function is calculated exactly as in the Abelian case. The only difference consists in the appearance of an additional factor, proportional to the trace of the product of the Γ-matrices at the vertex:

$$i(p+q)_\alpha\Gamma^{abc}_{\mu\nu\alpha}(p,q) = \text{const tr}\{\gamma_5[\Gamma_a,\Gamma_b]_+\Gamma_c\}\varepsilon_{\mu\nu\alpha\beta}p_\alpha q_\beta. \tag{11.25}$$

If the factor

$$A_{abc} = \mathrm{tr}\{\gamma_5[\Gamma_a, \Gamma_b]_+\Gamma_c\} \tag{11.26}$$

is not zero, then the function $\Gamma^{abc}_{\mu\nu\alpha}$ does not satisfy the generalized Ward identities, which leads to the loss of the gauge-invariance of the renormalized theory.

Let us analyze in which cases A_{abc} is equal to zero. For this, instead of the matrix Γ_a, we introduce the chiral matrices $T_\pm$

$$\Gamma_a = \frac{1}{2}(1+\gamma_5)T_a^+ \frac{1}{2}(1-\gamma_5)T_a^-, \tag{11.27}$$

where $T_\pm$ do not contain the matrix γ_5.

The factor A_{abc} can now be represented as

$$A_{abc} = 2(A^+_{abc} - A^-_{abc}), \tag{11.28}$$

where

$$A^\pm_{abc} = \mathrm{tr}\{[T_a^\pm, T_b^\pm]_+ T_c^\pm\}. \tag{11.29}$$

A_{abc} obviously becomes zero if $A^+ = A^-$. This is certainly fulfilled if the representations of $T_a^\pm$ are unitarily equivalent:

$$T_a^- = UT_a^+U^+, \tag{11.30}$$

where U is a unitary matrix. In this case, by another choice of the spinor-field basis, the interaction can be rewritten in a purely vector form:

$$\bar\psi\gamma_\mu\Gamma_a\psi = \frac{1}{2}\bar\psi\gamma_\mu\{(1+\gamma_5)T_a^+ + (1-\gamma_5)T_a^-\}\psi = \bar\psi'\gamma_\mu T_a^+\psi', \tag{11.31}$$

where

$$\psi' = \frac{1}{2}(1+\gamma_5)\psi + \frac{1}{2}(1-\gamma_5)U\psi. \tag{11.32}$$

Under such a redefinition of the fields ψ the γ_5-matrices appear in the mass terms. The absence of the anomalies in such models is absolutely natural. In the basis ψ' the gauge transformations no longer contain the matrices γ_5, and therefore one can apply to them the above described procedure of invariant regularization which allowed us to prove the generalized Ward identities rigorously. The actual form of the gauge group is not essential.

Such models are called "vectorlike," since at high energies, exceeding significantly all characteristic masses, they behave as models with purely vector interaction.

The unitary equivalence of T_+ and T_- is not necessary for the absence of anomalies. For this the fulfilment of the equality (11.28), which may be satisfied for other reasons, is sufficient.

Anomalies are absent also if $A^+_{abc} = A^-_{abc} = 0$, as occurs in some gauge groups. The sufficient condition for this is that: The representations realized by the matrices $T_a^\pm$ must be real. (A representation is called real if it is unitarily equivalent to its complex conjugate.) In this case,

$$\mathrm{tr}\{[T_a^{\pm}, T_b^{\pm}]_+ T_c^{\pm}\} = \mathrm{tr}\{[(T_a^{\pm})^*, (T_b^{\pm})^*]_+ (T_c^{\pm})^*\} = -\mathrm{tr}\{[T_a^{\pm}, T_b^{\pm}]_+ T_c^{\pm}\} = 0. \tag{11.33}$$

Such a situation is realized for the algebras

$$SU(2); \quad SO(2N+1), N \geq 2; \quad SO(4N), N \geq 2;$$
$$S_p(2N), N \geq 3;$$

$$G(2); \quad E(4); \quad E(7); \quad E(8), \tag{11.34}$$

all representations of which are real. (It can be shown that for algebras $SO(N)$ no anomalies are present for all $N \geq 5, N \neq 6$.) For the algebra $SU(3)$ anomalies are absent only for certain representations, for instance, 8 and $3 + \bar{3}$.

In non-Abelian theories one-loop diagrams with four external vector lines can also be anomalous. On the other hand, if in the given model all the one-loop diagrams satisfy the normal Ward identities, the multiloop diagrams are sure to be free from anomalies. This, as we have already pointed out, follows directly from the fact that the regularization by means of the higher covariant derivatives regularizes multiloop diagrams in any gauge theory, including those that contain γ_5-transformations. Therefore, multiloop diagrams automatically satisfy the normal Ward identities.

The above classification of "normal" and "anomalous" theories equally concerns theories with spontaneously broken symmetry. In the anomalous case the unitary and the renormalized gauges correspond to physically nonequivalent theories. In the unitary gauge, the theory is not renormalizable and therefore makes no sense in the framework of perturbation theory. On the contrary, in a renormalizable gauge the perturbation theory is constructed without difficulty; however, the S-matrix is not unitary in the space of physical states. Thus, requiring the absence of anomalies impose strict constraints on the possible gauge-invariant models.

Note that anomalies are not necessarily related to the violation of γ_5-invariance. In certain models, for example, anomalies arise that are related to the violation of conformal invariance.

In conclusion we shall discuss some general properties of non-Abelian γ_5-anomalies, which permit investigation of their structure without referring to perturbation theory.

As we already saw, anomalies arise in one-loop fermion diagrams involving γ_5-interaction with the vector field. Therefore we shall consider the generating functional

$$Z = \int \exp\left\{ i \int [i\bar{\psi}\gamma_\mu(\partial_\mu - V_\mu - A_\mu\gamma_5)\psi dx] \right\} d\bar{\psi} d\psi. \tag{11.35}$$

Here $V_\mu = V_\mu^a \Gamma_{kl}^a$ and $A_\mu = A_\mu^a \Gamma_{kl}^a$ are external classical sources for the vector and axial currents, respectively. In classical theory these currents are conserved owing to the invariance of the free fermion action with respect to global vector and axial transformations of the fermion fields:

$$\delta\psi_k = \Gamma^a_{kl}\beta^a\psi_l \tag{11.36}$$

$$\delta\psi_k = \Gamma^a_{kl}\alpha^a\gamma_5\psi_l.$$

The corresponding Ward identities assume

$$X_i Z = 0, \tag{11.37}$$

$$Y_i Z = G_i(V_\mu, A_\mu)Z. \tag{11.38}$$

Here

$$X^i = \partial_\mu \frac{\delta}{\delta V^i_\mu} - \left[V_\mu, \frac{\delta}{\delta V_\mu}\right]^i - \left[A_\mu, \frac{\delta}{\delta A_\mu}\right]^i, \tag{11.39}$$

$$Y^i = \partial_\mu \frac{\delta}{\delta A^i_\mu} - \left[A_\mu, \frac{\delta}{\delta V_\mu}\right]^i - \left[V_\mu, \frac{\delta}{\delta A_\mu}\right]^i, \tag{11.40}$$

and G_i represents an anomaly. In writing these identities we took advantage of the fact that the generating functional Z can always be defined so as to provide for conservation of the vector current $\bar{\psi}\gamma_\mu\Gamma\psi$.

The operators $X_i(x)$ and $Y_j(x')$ form an algebra;

$$[X_i(x), X_j(x')] = t^{ijk}\delta(x-x')X_k(x), \tag{11.41}$$

$$[X_i(x), Y_j(x')] = t^{ijk}\delta(x-x')Y_k(x), \tag{11.42}$$

$$[Y_i(x), Y_j(x')] = t^{ijk}\delta(x-x')X_k(x). \tag{11.43}$$

Applying the operators X_iY_j, X_iX_j, and Y_iY_j to the functional Z and taking advantage of the commutation relations (11.41) to (11.43) and the Ward identities (11.37) and (11.38), we obtain the conditions of compatibility the anomaly must satisfy:

$$X_i(x)G_j(x') = t_{ijk}\delta(x-x')G_k(x), \tag{11.44}$$

$$Y_i(x)G_j(x') - Y_j(x')G_i(x) = 0. \tag{11.45}$$

The first of these relations simply signifies that the anomaly G transforms by the adjoint representation of the vector gauge group. The second equation (11.45) represents the non-trivial condition of integrability that allows determination of the structure of the anomaly without performing calulations by perturbation theory. On the other hand, this condition can serve for checking whether the anomaly computed with the aid of some concrete regularization is correct. Thus, for example, the anomaly calculated by W. Bardeen for the case of the $SU(3)\times SU(3)$ group has the form

$$G_j = \frac{1}{8\pi^2}\varepsilon_{\mu\nu\sigma\rho}\mathrm{tr}\left[\frac{\lambda_j}{2}\left\{\frac{1}{4}V_{\mu\nu}V_{\sigma\rho} + \frac{1}{12}A_{\mu\nu}A_{\sigma\rho} + \frac{2}{3}i(A_\mu A_\nu V_{\sigma\rho} + A_\mu V_{\nu\sigma}A_\rho + V_{\mu\nu}A_\sigma A_\rho) - \frac{8}{3}A_\mu A_\nu A_\sigma A_\rho\right\}\right], \tag{11.46}$$

where

$$\begin{aligned} V_{\mu\nu} &= \partial_\mu V_\nu - \partial_\nu V_\mu - i[V_\mu, V_\nu] - i[A_\mu, A_\nu], \\ A_{\mu\nu} &= \partial_\mu A_\nu - \partial_\nu A_\mu - i[V_\mu, A_\nu] - i[A_\mu, V_\nu]. \end{aligned} \tag{11.47}$$

The anomaly (11.46) can be verified straightforwardly to satisfy the conditions of compatibility (11.44) and (11.45).

5

Some Applications and Conclusion

In this chapter we shall discuss possible applications of gauge theories to the description of elementary-particle interactions. At present the theory of gauge fields has been universally adopted as the theoretical basis of elementary particle physics, and a comprehensible review of its experimental applications would be far beyond the scope of this book. This issue is dealt with in special monographs, to which we refer the interested reader.

Here we shall restrict ourselves to the description of the most characteristic features of gauge-invariant elementary-particle models, without attempting to reflect the latest trends in this field. The examples to be considered are educational and illustrative in characters.

5.1 Unified Models of Weak and Electromagnetic Interactions

Until the late sixties electrodynamics was the only example of successful applications in elementary-particle physics of quantum field theory in general and gauge-invariant theories in particular. At the same time, it has been noticed for quite a while that weak and electromagnetic interactions have much in common. From experiments it is known that weak interactions involve vector currents. This leads to the idea that as in electrodynamics the interaction takes place through an exchange of vector particles, which have become known as intermediate bosons. Like the electromagnetic current, the weak current is conserved. And finally, the weak interaction is universal—the interaction is characterized by a single constant (if one neglects any effects due to mixing of various fundamental particles).

All these properties receive a natural explanation if one assumes that weak and electromagnetic interactions are described by a gauge-invariant theory, in which the Yang-Mills field plays the role of the interaction carrier. However, unlike the long-range electromagnetic interaction, the weak interaction has a finite interaction range, and consequently the corresponding vector fields must be massive. A second difference is that the weak interaction does not conserve parity. These differences, which for a long time hindered the construction of a realistic unified theory of weak and electromagnetic interactions, can be explained successfully by means of the Higgs mechanism. Spontaneous symmetry breaking allows one to select an "electromagnetic" direction in the internal charge space. The corresponding vector meson remains massless and interacts with the parity-conserving current. The other mesons acquire a nonzero mass, and their interaction does not conserve parity.

Let us consider the simplest realization of these ideas. The choice of the gauge group is, to a great extent, arbitrary. The dimension of the group must be not less than three, since it must involve, at a minimum, the generators corresponding to the photon (1) and the intermediate vector mesons (2). Taking into account only the "light" leptons—the electron, the muon, and the corresponding neutrinos—the minimum number of generators is equal to four. Indeed, the charged weak current has the following structure:

$$j_e^+ = \bar{\nu}_e \gamma_\mu (1 + \gamma_5) e = \bar{\psi}_e \gamma_\mu (1 + \gamma_5) \tau^+ \psi_e, \tag{1.1}$$

where

$$\psi_e = \begin{pmatrix} \nu_e \\ e \end{pmatrix} \tag{1.2}$$

(the muon current has an analogous structure).

Therefore the matrices $(1 + \gamma_5)\tau^+$ and $(1 + \gamma_5)\tau^-$ are involved in the Lie algebra of the gauge group. The minimal Lie algebra containing these matrices consists of the generator:

$$(1 + \gamma_5)\tau^+, \quad (1 + \gamma_5)\tau^-, \quad (1 + \gamma_5)\tau_3 \tag{1.3}$$

and corresponds to the $SU(2)$ group. This algebra does not contain a generator by means of which it could be possible to construct a parity-conserving electromagnetic current. The simplest algebra generating both the electromagnetic and the charged currents corresponds to the group $U(2)$ and contains four generators, one of which corresponds to the neutral weak current. Precisely this group forms a basis for the Weinberg-Salam model, which will be discussed below.

In the past, when the first unified models were being constructed, neutral currents were not known experimentally. Therefore, at that time alternative models were proposed which did not involve the interaction of neutral currents. Thus, for example, by introducing, besides the electron and muon, additional "heavy" leptons, G. Georgi and S. Glasho succeeded in constructing a gauge-invariant model based on the $SU(2)$ group. Subsequent experiments, however, confirmed the predictions of the Weinberg-Salam model concerning the existence of neutral currents. Moreover, it turned out that the heavy leptons discovered experimentally are also well described by the Weinberg-Salam model. All the data accumulated during the recent years testify unambigously in favor of this model. The ultimate triumph of the Weinberg-Salam model turned out to be direct experimental discovery of the neutral and of the charged intermediate vector mesons.

Below we shall describe in detail a model based on the group $SU(2) \times U(1)$— the Weinberg-Salam model. In the Weinberg-Salam model the electron and the electron neutrino are united in an SU_2 doublet L and singlet R

$$L = \frac{1}{2}(1+\gamma_5)\begin{pmatrix} \nu_e \\ e \end{pmatrix}, \quad R = \frac{1}{2}(1-\gamma_5)e. \tag{1.4}$$

This choice of the multiplets is due to the fact that the right-hand- and the left-hand-polarized leptons are involved in the weak interaction nonsymmetrically— the right-hand-polarized neutrino is not observed experimentally. The muon and the muon neutrino are united in analogous multiplets. Henceforth we shall limit ourselves to the consideration of the electron sector. By requiring the weak charged currents to have a V-A structure and the photon to interact only with the vector current of charged particles, we come to the following transformation laws

$$\begin{aligned} L(x) &\to L(x) - ig\frac{\tau^a}{2}\zeta^a(x)L(x) - \frac{ig_1}{2}\eta(x)L(x) + \ldots, \\ R(x) &\to R(x) - ig_1\eta(x)R(x) + \ldots \end{aligned} \tag{1.5}$$

Since the group $SU(2) \times U(1)$ is not simple, the gauge transformations involve two arbitrary parameters g and g_1. The following gauge fields correspond to the subgroups $SU(2)$ and $U(1)$: the isovector field A^a_μ and the singlet B_μ.

The gauge-invariant Lagrangian describing the interaction on the multiplets R and L with the Yang-Mills field has the form

$$\begin{aligned} \mathcal{L} = \frac{1}{8}\mathrm{tr}\mathcal{F}_{\mu\nu}\mathcal{F}_{\mu\nu} - \frac{1}{4}G_{\mu\nu}G_{\mu\nu} + \bar{L}i\gamma_\mu\left[\partial_\mu + ig\frac{\tau^a}{2}A^a_\mu + \frac{ig_1}{2}B_\mu\right]L \\ + \bar{R}i\gamma_\mu[\partial_\mu + ig_1B_\mu]R, \end{aligned} \tag{1.6}$$

where $\mathcal{F}_{\mu\nu}$ is the strength tensor of the Yang-Mills field, and $G_{\mu\nu}$ is the analogous tensor for the Abelian fields

$$G_{\mu\nu} = \partial_\nu B_\mu - \partial_\mu B_\nu \tag{1.7}$$

Note that the mass term

$$m(\bar{L}R + \bar{R}L) \tag{1.8}$$

for leptons is forbidden by the requirement of invariance under the transformations (1.5).

All the fields involved inthe Lagrangian (1.6) have zero masses. However, if the fields $\mathcal{A}_\mu, B_\mu$ and R, L interact also with scalar fields, then they may also acquire nonzero masses due to the Higgs effect. Since all vector mesons, with the exception of the photon, must become massive, we shall take advantage of the concrete spontaneous-symmetry-breaking model (3.25) of Chapter 1. We introduce the complex double.

$$\varphi = \begin{pmatrix} \varphi_1 \\ \varphi_2 \end{pmatrix} \tag{1.9}$$

which transforms under the gauge transformations in the following ways:

$$\varphi \to \varphi - ig\zeta^a(x)\frac{\tau^a}{2}\varphi + \frac{ig_1}{2}\eta(x)\varphi(x). \tag{1.10}$$

The gauge-invariant Lagrangian describing the interaction with the fields $\mathcal{A}_\mu, B_\mu, R, L$ has the form

$$\mathcal{L} = \left|\partial_\mu\varphi + ig\frac{\tau^a}{2}A^a_\mu\varphi - \frac{ig_1}{2}B_\mu\varphi\right|^2$$
$$-G\{(\bar{L}\varphi)R + \bar{R}(\varphi^+L)\} + \frac{m^2}{2}(\varphi^+\varphi) - \lambda^2(\varphi^+\varphi)^2 \tag{1.11}$$

As we already know, an interaction of the type (1.11) generates spontaneous symmetry breaking: The vacuum expectation value of the field φ differs from zero, and to construct the perturbation theory around the asymmetric ground state it is necessary to pass to the shifted fields

$$\varphi \to \varphi' = \begin{pmatrix} \varphi_1 \\ \varphi_2 + \mu \end{pmatrix}; \qquad (\mathrm{Im}\mu = 0). \tag{1.12}$$

As a result of this shift, mass terms for the vector fields appears:

$$\frac{\mu^2}{4}[g^2(A^1_\mu)^2 + g^2(A^2_\mu)^2 + (g_1B_\mu - gA^3_\mu)^2]. \tag{1.13}$$

The diagonalization of the quadratic form (1.13) leads to the following spectrum of masses.

The charged mesons $W_\pm$,

$$W_{\mu\mp} = \frac{A^1_\mu \pm A^2_\mu}{\sqrt{2}} \tag{1.14}$$

acquire masses

$$m_W = \frac{1}{\sqrt{2}}g\mu. \tag{1.15}$$

The neutral mesons

$$Z_\mu = (g^2 + g_1^2)^{-1/2}(-gA^3_\mu + g_1B_\mu) \tag{1.16}$$

and

$$A_\mu = (g^2 + g_1^2)^{-1/2}(g_1A^3_\mu + gB_\mu) \tag{1.17}$$

acquire the masses $(\mu/\sqrt{2})(g^2+g_1^2)^{1/2}$ and 0, respectively. As a result of the shift (1.12), the leptons also acquire nonzero masses. The mass term has the form

$$-G\left\{\bar{L}\begin{pmatrix}0\\ \mu\end{pmatrix}R+\bar{R}(0,\mu)L\right\}=-G\mu\bar{e}e. \tag{1.18}$$

The neutrino remains massless.

Finally, using the expansion

$$\varphi_1=\frac{1}{\sqrt{2}}(iB_1+B_2);\quad \varphi_2=\mu+\frac{1}{\sqrt{2}}(\sigma-iB_3), \tag{1.19}$$

we find that the field σ acquires a mass $2\lambda\mu$.

The Goldstone fields B_i have a zero mass and, as usual, can be removed by a gauge transformation.

The interaction of the leptons with vector fields has the form

$$\begin{aligned}\mathcal{L}_I=\frac{-g}{2\sqrt{2}}\bar{\nu}_e\gamma_\mu(1+\gamma_5)eW_\mu^+ + \text{e. c.}+\frac{gg_1}{(g^2+g_1^2)^{1/2}}\bar{e}\gamma_\mu eA_\mu\\ +\frac{e}{2}[\text{tg}\theta_W(2\bar{e}_R\gamma^\mu e_R+\bar{\nu}_e\gamma^\mu\nu_e+\bar{e}_L\gamma^\mu e_L)\\ -\text{ctg}\theta_W(\bar{e}_L\gamma^\mu e_L-\bar{\nu}_e\gamma^\mu\nu_e)]Z_\mu+\{e\leftrightarrow\mu,\nu_e\leftrightarrow\nu_\mu\},\end{aligned} \tag{1.20}$$

where the quantity $\theta_W=\text{arctg}(g'g^{-1})$ is called the Weinberg angle. From this formula it is seen that the electromagnetic constant e and the Fermi weak-interaction constant G_F are expressed in terms of the parameters g and g_1 in the following ways

$$e=\frac{gg_1}{\sqrt{g^2+g_1^2}}=g\sin\theta_W, \tag{1.21}$$

$$\frac{G_F}{\sqrt{2}}=\frac{g^2}{8m_W^2}. \tag{1.22}$$

From (1.21) it follows that

$$e\leq g, \tag{1.23}$$

whence

$$m_W=\left(\frac{g^2\sqrt{2}}{8G_F}\right)^{1/2}\gtrsim\frac{38}{\sin\theta_W}\text{ GeV}\geq 38\text{ GeV} \tag{1.24}$$

that is, the mass of the charged intermediate meson has a lower bound and is large.

An analogous estimate of the neutral meson gives

$$m_Z\geq 75\text{ GeV} \tag{1.25}$$

The Weinberg-Salam model itself yields only the lower limits for the masses of the intermediate mesons. The Weinberg angle, however, can be determined from studies of processes involving weak neutral currents. The present experimental values are $\sin^2\theta_W \approx 0.218$. Substituting this value into the formula (1.24), we find for the mass of the charged vector meson $m_W \simeq 80$ GeV, and for the neutral meson $m_Z \simeq 90$ GeV. The experimental masses $m_W = 81.8 \pm 1.5$ GeV and $m_Z = 92.6 \pm 1.7$ GeV are in excellent agreement with the predictions of the model.

Besides the terms written above, the interaction Lagrangian also describes the self-interaction of scalar mesons and their interaction with leptons. The existence and properties of the Higgs scalar mesons represent the least clear issue from an experimental point of view. Within the framework of the Weinberg-Salam model the mass of the σ-meson is a free parameter, but it cannot, however, be made indefinitely large. This is evident, for instance, from the fact that in the limit $m_\sigma \to \infty$ the amplitudes, to which diagrams with internal σ-lines correspond, tend toward infinity. Indirect estimates of the mass of the σ-meson exist which do not contradict experimental data. Nevertheless, attempts at direct observation of Higgs mesons have hitherto been unsuccessful, and this issue still remains open.

As we have already pointed out, the most interesting prediction of the Weinberg-Salam model is the existence of neutral currents. It leads to a whole series of experimental consequences concerning, for instance, elastic e-ν scattering processes and parity non-conservation effects in atomic physics. All these predictions have been confirmed.

What concerns the interaction of charged weak current, here the predictions of the Weinberg-Salam model coincide, in the lowest order, with the predictions of the phenomenological four-fermion model. But, because, unlike the latter, the Weinberg-Salam model is renormalizable, it also permits the calculation of radiational corrections of higher orders. Thus, for instance, the weak corrections to the anomalous magnetic moment of the muon turn out to be small, in agreement with experimental data.

Finally, the results of studies of the heavy τ-lepton ($m \simeq 2$ GeV$/c^2$) and of the respective neutrino are also compatible with the framework of the Weinberg-Salam model, if the τ-lepton is considered to be, as the muon, a heavy copy of the electron, i.e., its interaction is considered to be of a form similar to (1.20).

Let us discuss in more detail the renormalization of the Weinberg-Salam model. Since the Lagrangian (1.6), (1.11) is gauge-invariant, the renormalization procedure described in the preceding chapter can be applied to it. However, the gauge group contains the Abelian subgroup $U(1)$ and, in accordance with the classification of Section 4.11, is anomalous. Therefore, although formally gauge-invariant, the Weinberg-Salam model described by the Lagrangian (1.6), (1.11) is not renormalizable. The situation may be improved by means of the mechanism described in the preceding chapter. As we saw, anomalies are absent in the case of any gauge group, if the right-hand- and the left-hand-polarized fermions give to the anomalous triangle diagram contributions of equal magnitudes and opposite signs. Therefore, the introduction in the Weinberg-Salam model, in addition to the electron multiplets

(1.1), of multiplets with opposite helicities

$$\tilde{R} = \frac{1}{2}(1-\gamma_5)\begin{pmatrix} N \\ E \end{pmatrix}, \qquad \tilde{L} = \frac{1}{2}(1+\gamma_5)E, \tag{1.26}$$

interaction with vector fields in the same way as L and R, leads to the absence of anomalies in such a modified model. Note, however, that for the $\tilde{L}$ and $\tilde{R}$ it is not possible to use the muon and muon neutrino, since the "compensating" leptons must be involved in the weak interaction with opposite helicities.

Therefore, a renormalizable extension of the Weinberg-Salam model can be constructed within the framework of lepton physics only if heavy heptons are introduced with helicities opposite to that of the electron. Such interactions have not been observed.

Luckily, the possibility remains of compensating the lepton anomalies at the expense of similar anomalies in electroweak interactions of hadrons.

Let us now pass to the discussion of the weak and electromagnetic interactions of hadrons. Similarly to the Weinberg-Salam model, which predicts the existence of neutral lepton currents, the analogous model for the hadron sector predicts the existence of weak neutral hadron currents. If the hadron symmetry group is the group $SU(3)$, then the neutral current contains strangeness-changing terms. To verify this we recall that in the $SU(3)$-symmetric theory the weak charged hadron current is described by the Cabbibo formula

$$j_\mu^+ = \bar{u}\gamma_\mu(1+\gamma_5)(d\cos\theta + s\sin\theta). \tag{1.27}$$

Here u and d are the (up) and (down) quarks with charges $2/3$ and $-1/3$, s is the strange quark with a charge $-1/3$, and θ is the Cabbibo angle, characterizing the relative probabilities of processes with and without a change of strangeness.

As in the Weinberg-Salam model for leptons, the generators corresponding to the charged currents generate the group $SU(2)$. Therefore the gauge-invariant theory involves, together with the charged currents (1.27), a neutral current of the form

$$j_\mu^3 = \bar{u}\gamma_\mu(1+\gamma_5)u + (\bar{d}\cos\theta + \bar{s}\sin\theta)\gamma_\mu(1+\gamma_5)(d\cos\theta + s\sin\theta). \tag{1.28}$$

The current j_μ^3 interacts with the third component of the Yang-Mills field A_μ^3 and consequently represents a linear combination of the electromagnetic and weak charged currents. As a result, such a model allows processes involving neutral strangeness-changing currents, such as

$$K_L^0 \to \mu^+\mu^-, \qquad K^+ \to \pi^+\nu\bar{\nu}, \tag{1.29}$$

and the probabilities of these processes should be comparable to the probabilities of processes involving charged currents. From experiments it is known that processes such as (1.29) are forbidden to a very high degree of accuracy. The ratio of the decay probability for $K_L^0 \to \mu^+\mu^-$ to the probability of the decay $K^+ \to \mu^+\nu_\mu$, involving charged currents, is $\leq 10^{-9}$. It is possible to forbid such processes in the gauge-invariant theory by giving up the assumption of the $SU(3)$ structure of hadrons. The simplest possibility is to substitute the group $SU(4)$ for the group

$SU(3)$. In the quark model this is equivalent to the introduction of the fourth quark c with a new quantum number—the "charm."

The four-quark gauge model of weak and electromagnetic interactions is constructed in the same way as the Weinberg-Salam model for leptons. The left-hand-polarized quarks are united in two $SU(2)$ doublets

$$L_1 = \frac{1}{2}(1+\gamma_5)\begin{pmatrix} u \\ d\cos\theta + s\sin\theta \end{pmatrix};$$

$$L_2 = \frac{1}{2}(1+\gamma_5)\begin{pmatrix} c \\ -d\sin\theta + s\cos\theta \end{pmatrix}, \tag{1.30}$$

and the right-hand-polarized ones in the singlets

$$R_1 = \frac{1}{2}(1-\gamma_5)u, \qquad R_2 = \frac{1}{2}(1-\gamma_5)c,$$

$$R_3 = \frac{1}{2}(1-\gamma_5)(d\cos\theta + s\sin\theta);$$

$$R_4 = \frac{1}{2}(1-\gamma_5)(-d\sin\theta + s\cos\theta). \tag{1.31}$$

The charged hadron current has the form

$$j_u^+ = \bar{u}_L\gamma_\mu(d\cos\theta + s\sin\theta) + \bar{c}_L\gamma_\mu(-d\sin\theta s\cos\theta). \tag{1.32}$$

Commuting j_μ^+ and j_μ^-, we obtain for j_μ^3 the expression

$$j_\mu^3(x) = \left[\int j_0^+(y)d^3y, j_\mu^-(x)\right]$$

$$= \bar{u}\gamma_\mu(1+\gamma_5)u + \bar{d}\gamma_\mu(1+\gamma_5)d + \bar{s}\gamma_\mu(1+\gamma_5)s. \tag{1.33}$$

In this current, strangeness-changing terms are absent, and consequently processes such as (1.29) are forbidden in the lowest order of the weak interaction.

We shall not write out here the total gauge-invariant Lagrangian for the weak and electromagnetic interactions of hadrons. It is entirely analogous to the Lagrangian (1.3). Its most remarkable feature is the prediction of "charmed" hadron states. Recent experiments have confirmed this prediction of the gauge theories also.

To complete the description of the models of weak hadron interactions we recall that, according to the widely accepted point of view, three varieties of quarks exist, differing from one another by their "color;" that is each quark p, c, d, s can assume three different "colors." Weak interactions are not sensitive to the color, and the corresponding Lagrangian is a sum of three identical Lagrangians. The hypothesis of the existence of color was put forward in order to explain the observed spectrum of hadrons within the framework of the usual assumptions on the relationship between the spin and statistics. Remarkably, it turned out that introduction of color at the same time makes the unified model of weak interactions, described above, self-consistent. In the model involving four leptons μ, e, ν_μ, ν_e and four quarks of three colors p, c, d, s with charges $2/3, 2/3, -1/3, -1/3$, anomalies are absent; and consequently the corresponding theory is renormalizable. In this case the total lepton charge (-2) is equal in magnitude and opposite in sign to the total charge of the quarks $(2/3 \times 3 = 2)$, and for this reason the anomalies of the lepton and hadron currents are compensated.

An attentive reader may notice that we have now only demonstrated the compensation of electron and muon anomalies. But above we mentioned that at present one more lepton family is known, the τ-lepton and the respective neutrino. Since these particles interact precisely like the electron and the muon, they also contribute to the triangular anomaly. To compensate this contribution it is necessary to introduce two more quarks termed the t (top) and the b (bottom) quarks. Hadrons including the b quark have already been discoverd, and the search for the t quark is still underway.

At present we cannot say whether new leptons and quarks will be observed and whether their interaction will also satisfy the condition for compensation of anomalies. However, all presently available data are successfully explained within the framework of the renormalizable gauge invariant Weinberg-Salam model.

5.2 Asymptotic Freedom. Gauge Theories of Strong Interactions

At first sight the dynamics of strong interactions seems to be too complicated to try to describe it in the framework of any reasonable quantum-field-theory model. Until recently, for the description of strong interactions either dispersion-relation methods, based on the most general physical requirements of causality and unitarity, or phenomenological models have been used. Attempts to construct a relativistic Lagrange model, which would give a detailed description of the dynamics of strong interactions, have not led even to qualitative results.

On the other hand, deep-inelastic lepton-proton scattering experiments yield evidence that a simple dynamical mechanism forms the basis of strong interactions. At large momentum transfers, which are equivalent to small spatial distances, hadron behave as if they consisted of noninteracting point objects. Thus, the following qualitative picture arises: Hadrons are composite objects, and the interaction between their components tends to zero at small distances. At the same time, at large distances the effective interaction becomes strong, so that a hadron is a strongly bound system.

Is it possible to describe such an interaction in the framework of any quantum-field-theory model? The answer to this question turns out to be unambiguous. The above-described behavior of the interaction can be obtained only in a non-Abelian gauge theory. All consistent field-theory models that do not involve the Yang-Mills field lead to an increase of the effective interaction at small distances. This unique feature of the Yang-Mills fields is due to the phenomenon of asymptotic freedom, to the description of which we shall now pass.

We shall now discuss the asymptotic behavior of the Green functions in the deep-Euclidean region where the squares of all the momentum arguments p_i are negative and have large absolute values. This asymptotic behavior has no direct physical meaning, of course, since for the calculation of the S-matrix the values of the Green functions at $p_i^2 = m_i^2 \geq 0$ are needed. However, it may be shown that the probabilities of deep-inelastic scattering processes are directly related to the behavior of the Green functions in the deep-Euclidean region.

To be more precise, we shall investigate the asymptotic behavior of strongly connected proper vertex functions $\Gamma_n(\mathcal{N}p_1, \ldots \mathcal{N}p_n, m, g)$, where $p_i^2 = -a_i^2 < 0$ as $\mathcal{N} \to \infty$. For this we shall need the technique of the renormalization group, the main concepts of which we shall briefly recall.

As we already know, the subtraction of the leading terms of the Taylor expansion of divergent proper vertex functions is equivalent to the insertion in the Lagrangian of local counterterms, which in turn is equivalent to the renormalization of the parameters involved in the Lagrangian. The transition from one subtraction point to another is equivalent to a finite renormalization. For instance, the insertion of the counterterms

$$\frac{1}{8}\operatorname{tr}\{(z_2-1)(\partial_\nu \mathcal{A}_\mu - \partial_\mu \mathcal{A}_\nu)^2 + 2g(z_1-1)(\partial_\nu \mathcal{A}_\mu - \partial_\mu \mathcal{A}_\nu)[\mathcal{A}_\mu, \mathcal{A}_\nu] + (z_1^2 z_2^{-1} - 1)[\mathcal{A}_\mu, \mathcal{A}_\nu]^2\} \quad (2.1)$$

(where ... stands for the corresponding counterterms for fictitious particles and the fields of matter) is equivalent to the following renormalizations of the Green functions and the charges:

$$\begin{aligned} G^{\mathrm{tr}}_{\mu\nu}(k,g) &\to z_2^{-1} G^{\mathrm{tr}}_{\mu\nu}(k,g'), \\ \Gamma_{A^3}(p,q,g) &\to z_1 \Gamma_{A^3}(p,q,g'), \\ \Gamma_{A^4}(k,p,q,g) &\to z_1^2 z_2^{-1} \Gamma_{A^4}(k,p,q,g'), \\ g \to g' &= z_1 z_2^{-3/2} g. \end{aligned} \quad (2.2)$$

Now we introduce a dimensionless function D related to $G^{\mathrm{tr}}_{\mu\nu}$ by the relation (for simplicity we shall further work in the Lorentz gauge)

$$G^{\mathrm{tr}}_{\mu\nu} = \left(g^{\mu\nu} - \frac{k^\mu k^\nu}{k^2}\right)\frac{D}{k^2} \quad (2.3)$$

The dimensionless functions, obtained upon the singling out tensor structures from Γ_{A_3} and Γ_{A_4} of tensor structures, will be denoted as $g\Gamma_3$ and $g^2\Gamma_4$. These functions depend only on dimensionless arguments:

$$D = D\left(\frac{k_1^2}{\lambda}, \frac{m^2}{\lambda}, g\right); \qquad \Gamma_3 = \Gamma_3\left(\frac{k_2^2}{\lambda}, \ldots, \frac{k_4^2}{\lambda}, \frac{m^2}{\lambda}, g\right),$$

$$\Gamma_4 = \Gamma_4\left(\frac{k_5^2}{\lambda}, \ldots, \frac{k_{11}^2}{\lambda}, \frac{m^2}{\lambda}, g\right) \tag{2.4}$$

$$k_2^2 \equiv p^2, \quad k_3^2 \equiv q^2, \quad k_4^2 = (p+q)^2,$$

λ is the subtraction point. (The invariant variables are chosen so as to provide for the functions Γ_i being real for $k_i^2 = \lambda < 0$.)

We shall consider the functions D, Γ_3, Γ_4 to be normalized by the condition

$$D, \Gamma_3, \Gamma_4 = 1 \qquad \text{for} \qquad \frac{k_i^2}{\lambda} = 1. \tag{2.5}$$

In the tree approximation all these functions are obviously equal to unity. Account of radiative corrections leads to a non-trivial dependence on the momenta; however, at the normalization point this dependence vanishes, and once again we return to the quasi-classical limit. This means that the charge g characterizes the effective interaction in the vicinity of the normalization point. If subtraction is performed near the zero point (as usually it is in electrodynamics), then the constant g characterizes the low-energy interaction. Contrariwise, subtraction at $k^2 = \lambda \to -\infty$ corresponds to the effective interaction constant in the deep-Euclidean region.

Transition from one subtraction point to another in accordance with (2.2) is equivalent to the following renormalization of the charge:

$$g \to g' = z_1 z_3^{-3/2} g. \tag{2.6}$$

Here the factor z_1 is due to renormalization of the triple vetex, and the factor $z_2^{-3/2}$ is caused by three propagators converging at this vertex, each one of them connecting two vertices (Fig. 29).

We shall introduce the function $\bar{g}(\frac{k^2}{\lambda}, \frac{m^2}{\lambda}, g)$ called the invariant, or the running, charge:

$$\bar{g}\left(\frac{k^2}{\lambda}, \frac{m^2}{\lambda}, g\right) = g\Gamma_3\left(\frac{k^2}{\lambda}, \frac{k^2}{\lambda}, \frac{m^2}{\lambda}, g\right) \times \left[D\left(\frac{k^2}{\lambda}, \frac{m^2}{\lambda}, g\right)\right]^{3/2}. \tag{2.7}$$

If together with the transformation of the Green functions (2.2) we perform the following compensating transformation of the charge:

$$g \to z_1^{-1} z_2^{3/2} g, \tag{2.8}$$

then, clearly, the function $\bar{g}$ will not undergo any change.

The transformation (2.2) and (2.8) form a group of multiplicative renormalizations, or a renormgroup, and the running charge $\bar{g}$ represents an invariant of this group.

The condition for the theory to be independent of the choice of the subtraction point can be written, together with the compensating transformation of the charge,

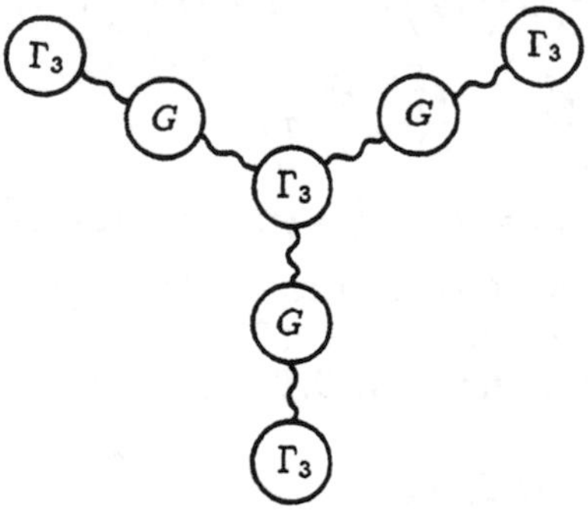

Figure 29.

(renorm-invariance) in the form

$$D\left(\frac{k^2}{\lambda_2},\frac{m^2}{\lambda_2},g_2\right)=z_2D\left(\frac{k^2}{\lambda_1},\frac{m^2}{\lambda_1},g_1\right),$$

$$\Gamma_3\left(\frac{k_1^2}{\lambda_2}\cdots\frac{k_3^2}{\lambda_2},\frac{m^2}{\lambda_2},g_2\right)=z_1^{-1}\Gamma_3\left(\frac{k_1^2}{\lambda_1}\cdots\frac{k_3^2}{\lambda_1},\frac{m^2}{\lambda_1},g_1\right),$$

$$\Gamma_4\left(\frac{k_4^2}{\lambda_2}\cdots\frac{k_{10}^2}{\lambda_2},\frac{m^2}{\lambda_2},g_2\right)=z_1^{-2}z_2^{\frac{1}{2}}\Gamma_4\left(\frac{k_4^2}{\lambda_1}\cdots\frac{k_{10}^2}{\lambda_1},\frac{m^2}{\lambda_1},g_1\right),$$

$$g_2=z_1z_2^{-3/2}g_1. \tag{2.9}$$

From the equations (2.4) and the normalization condition it follows that

$$z_2=D\left(\frac{\lambda_1}{\lambda_2},\frac{m^2}{\lambda_2},g_2\right),$$

$$z_1^{-1}=\Gamma_3\left(\frac{\lambda_1}{\lambda_2},\frac{\lambda_1}{\lambda_2},\frac{\lambda_1}{\lambda_2},\frac{m^2}{\lambda_2},g_2\right). \tag{2.10}$$

Using the formulas (2.7) and (2.10), we can write the condition of invariance of the running charge in the form

$$\bar g\left(\frac{k^2}{\lambda_1},\frac{m^2}{\lambda_1},g\right)=\bar g\left(\frac{k^2}{\lambda_2},\frac{m^2}{\lambda_2},\bar g\left(\frac{\lambda_2}{\lambda_1},\bar g\right)\right) \tag{2.11}$$

We shall consider this condition in the deep-Euclidean region, $k^2\to-\infty$, and choose the subtraction point so that $|\lambda|\gg m^2$. It can be shown that in renormalizable theories the leading terms in the asymptotics of the Green functions are independent of the mass, and therefore we may set $m = 0$ in the equation (2.11). Then the condition of invariance assumes the form

$$\bar{g}(\mathcal{N}, g) = \bar{g}\left(\frac{\mathcal{N}}{t}, \bar{g}(t, g)\right). \tag{2.12}$$

This equation indicates that scaling of the momenta by a factor of t is equivalent to transition from the coupling constant g to the new constant $\bar{g}(t, g)$. Therefore knowledge of the behavior of the running charge in the vicinity of the normalization point makes possible investigation of its behavior at high Euclidean momenta. To this end we differentiate the equation (2.12) with respect to $\mathcal{N}$ and set $\mathcal{N} = t$. Thus we obtain

$$\mathcal{N}\frac{\partial \bar{g}}{\partial \mathcal{N}} = \beta(\bar{g}), \tag{2.13}$$

where

$$\beta(g) = \left.\frac{\partial \bar{g}(t, g)}{\partial t}\right|_{t=1}. \tag{2.14}$$

The function $\beta(g)$ is called the Gell-Mann-Low function. It can be seen to depend only on the behavior of the invariant charge in the vicinity of the normalization point.

The equation (2.14) is conveniently rewritten in the integral form

$$\int_g^{\bar{g}(\mathcal{N}, g)} \frac{da}{\beta(a)} = \ln \mathcal{N}, \tag{2.15}$$

where the following boundary condition is taken into account:

$$\bar{g}(1, g) = g. \tag{2.16}$$

As $\mathcal{N} \to \infty$, the right-hand side of (2.15) tends to infinity. This means that in the limit $\mathcal{N} \to \infty$ the invariant charge either increases indefinitely or tends toward a constant value that is a zero of the β-function. The points at which the β-function equals zero are called fixed points. If within the vicinity of a fixed point g_0 the derivative of the β-function is negative, then such a point is said to be ultraviolet-stable. In this case $\bar{g}(\mathcal{N}) \to g_0$ as $\mathcal{N} \to \infty$. If the derivative of the β-function in the vicinity of g_0 is positive, then such a point is called infrared-stable. In this case $\bar{g}(\mathcal{N}) \to g_0$ as $\mathcal{N} \to 0$.

The behavior of the β-function in the vicinity of the point $g_0 = 0$ is of special interest, since in practice the only reliable way of calculating the β-function is application of returbation theory. Here two essentially differing cases are possible, they are graphically depicted in Fig. 30.

In the first case (Fig. 30, a) the zero is an infrared-stable point; therefore when $\mathcal{N}$ is small, the effective coupling constant is small, and the perturbation theory is applicable. In the opposite limit $\mathcal{N} \to \infty$ the effective constant increases and departs from zero. In doing so it may either increase indefinitely or, as shown in Fig. 30, a, tend toward the ultraviolet-stable point.

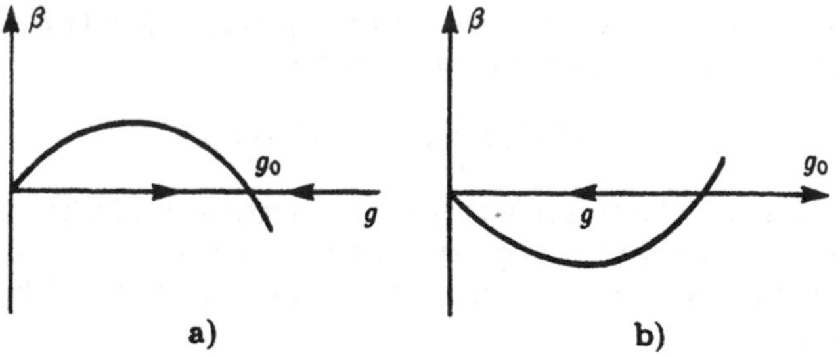

Figure 30. Various types of fixed points.

In the second case (Fig. 30, b) the zero is an ultraviolet-stable point. When $\mathcal{N} \to \infty, \bar{g}(\mathcal{N}) \to 0$. Therefore in the deep-Euclidean region we are justified in applying perturbation theory. Such theories are called asymptotically free since at small distances the interaction tends toward zero.

In most quantum-field-theory models a second possibility is realized. For example, in electrodynamics in the lowest order in α,

$$\beta(\alpha) = \frac{\alpha^2}{3\pi}. \tag{2.17}$$

Substituting this value into the formula (2.18), we obtain

$$\bar{\alpha}(\mathcal{N}, \alpha) = \frac{\alpha}{1 - \frac{\alpha}{3\pi} \ln x}. \tag{2.18}$$

As is seen, as $\mathcal{N}$ increases, $\bar{\alpha}(\mathcal{N}, \alpha)$ increases, and at $\mathcal{N} = e^{3\pi/\alpha}$ it goes to infinity. Of course, in reality at $\mathcal{N} \sim e^{3\pi/\alpha}$ the formula (2.18) cannot be used, since the function β has been calculated assuming the effective coupling constant to be small.

If we nevertheless try to extrapolate the formula (2.18) to the region of large $\bar{\alpha}$, we immediately come to a contradiction. In the electrodynamics the invariant charge is related to the photon Green function by

$$\bar{\alpha}(\mathcal{N}, \alpha) = \alpha d(\mathcal{N}, \alpha), \tag{2.19}$$

where

$$D_{\mu\nu}(k) = -\frac{1}{k^2}\left(g_{\mu\nu} - \frac{k_\mu k_\nu}{k^2}\right) d(k^2, \alpha). \tag{2.20}$$

Therefore, if the denominator in the expression (2.18) becomes equal to zero, that means the photon Green function has a pole. It is not difficult to verify that the residue at this pole is negative. The corresponding state has a negative norm, which is incompatible with the unitarity condition. Thus, in the case of $\beta(g) > 0$ at $g \sim 0$, the perturbation theory cannot give any realiable information about the asymptotic behavior of the Green functions.

The case of the Yang-Mills theory is quite different. In this theory $\beta(g)$ is negative in the vicinity of zero, and consequently zero is an ultraviolet-stable point.

Indeed, by definition,

$$\beta(g) = \left.\frac{\partial \bar{g}(\mathcal{N}, g)}{\partial \mathcal{N}}\right|_{\mathcal{N}=1} \tag{2.21}$$

where in the case of the Yang-Mills field the invariant charge is equal to

$$\bar{g}^2(\mathcal{N}, g) = g^2 \Gamma_3^2 D^3(\mathcal{N}). \tag{2.22}$$

Since $\mathcal{N} = k^2/\lambda$,

$$\left.\frac{\partial}{\partial \mathcal{N}}\right|_{\mathcal{N}=1} = -\lambda \left.\frac{\partial}{\partial \lambda}\right|_{\lambda=k^2} = -\left.\frac{\partial}{\partial \ln \lambda}\right|_{\lambda=k^2}. \tag{2.23}$$

On the other hand,

$$\left.\frac{\partial}{\partial \ln \lambda} D\left(\frac{k^2}{\lambda}\right)\right|_{\lambda=k^2} = -\left.\frac{\partial}{\partial \ln \lambda} z_2^{-1}\left(\frac{\Lambda}{\lambda}\right)\right|_{\lambda=\Lambda}, \tag{2.24}$$

$$\left.\frac{\partial}{\partial \ln \lambda} \Gamma_3\left(\frac{k^2}{\lambda}\right)\right|_{\lambda=k^2} = -\left.\frac{\partial}{\partial \ln \lambda} z_1\left(\frac{\Lambda}{\lambda}\right)\right|_{\lambda=\Lambda}. \tag{2.25}$$

Therefore, for the definition of $\beta(g)$ we can use the values of z_i determined previously. Thus we obtain

$$\beta(g^2) = -\frac{22}{3}\frac{g^4}{(4\pi)^2}. \tag{2.26}$$

Consequently, the square of the invariant charge tends to zero as $\mathcal{N} \to \infty$:

$$\bar{g}^2(\mathcal{N}, g^2) = \frac{1}{1 + \frac{g^4}{(4\pi)^2}\frac{22}{3}\ln \mathcal{N}}. \tag{2.27}$$

In the deep-Euclidean region the interaction "dies out," and the theory behaves as free. For the case of an arbitrary gauge group, and taking into account the interaction with the fields of matter, the function $\beta(g)$ is given by the formula

$$\beta(g^2) = \left[-\frac{11}{3}C(G) + \frac{4}{3}T(R)\right]\frac{g^2}{16\pi^2}, \tag{2.28}$$

$$\delta^{ab}C(G) = t^{acd}t^{bcd}; \qquad T(R)\delta^{ab} = \mathrm{tr}\{\Gamma^a, \Gamma^b\}. \tag{2.29}$$

where t^{acd} are the structure constants of the group and Γ^i are the generators of the representation realized by the fields of matter. If the number of multiplets of the fields of matter is not too large, then in this case also the theory is asymptotically free.

Thus, if the Yang-Mills fields are the carriers of the strong interactions, then at small distances quasifree particles will indeed be observed, in agreement with deep-inelastic scattering experiments.

Hitherto we considered the asymptotic behavior of only the invariant charge. We shall now study the behavior of the other Green functions. In principle, this can be done with the aid of the equations (2.9). We shall, however, make use of an alternative approach that will permit us to become familiar with one more renormgroup equation. Consider an arbitrary n-legged connected Green function. The condition of renorm-invariance for such a function (analog of the equations (2.9)) has the form

$$G_{(n)}(k_1^2 \ldots k_n^2, \lambda_0, g) = z_2^{n/2}\left(\frac{\lambda}{\lambda_0}, g\right) G_{(n)}(k_1^2 \ldots k_n^2, \lambda, \bar{g}), \tag{2.30}$$

$$\bar{g} = z_1 z_2^{-3/2} g.$$

(As before, we are working in the deep-Euclidean region, and therefore we omit mass terms). Here z_2 is the constant of finite renormalization of external lines corresponding to transition from the normalization point λ_0 to the point λ. Since the left-hand side of (2.30) is independent of λ, we can equate to zero the derivative with respect to λ of the right-hand side of this equation. Performing differentiation and assuming $\lambda_0 = \lambda$, we obtain

$$\frac{n}{2}\frac{\partial \ln z_2}{\partial \lambda} \cdot G_{(n)} + \frac{\partial G_{(n)}}{\partial \lambda} + \frac{\partial \bar{g}}{\partial \lambda}\frac{\partial G_n}{\partial \bar{g}}\bigg|_{\lambda=\lambda_0} . \tag{2.31}$$

Introducing the function

$$\gamma(\lambda) = \frac{1}{2}\lambda\frac{\partial}{\partial \lambda}\ln z_2\left(\frac{\lambda_0}{\lambda}, g\right)\bigg|_{\lambda_0=\lambda} \tag{2.32}$$

which is called the anomalous dimension, and recalling that

$$\beta(g) = \frac{\partial \bar{g}(t, g)}{\partial t}\bigg|_{t=1} = \frac{\partial \bar{g}\left(\frac{\lambda}{\lambda_0}, g\right)}{\partial \lambda}\lambda\bigg|_{\lambda=\lambda_0=0}, \tag{2.33}$$

we can write the equation (2.31) in the form

$$\left(\lambda\frac{\partial}{\partial \lambda} + \beta(g)\frac{\partial}{\partial g} + n\gamma(\lambda)\right) G_{(n)}(k_1 \ldots k_n, \lambda, g) = 0 \tag{2.34}$$

This equation, called the Callan-Symanzik equation, is convenient for investigation of the asymptotic behavior of the Green functions under simultaneous rescaling of all momenta. Since the charge g is a function of the subtraction point λ, we can in the equation (2.34) set

$$\lambda\frac{\partial}{\partial \lambda} + \beta(g)\frac{\partial}{\partial g} = \lambda\frac{d}{d\lambda} \tag{2.35}$$

and write the Callan-Symanzik in the form

$$\left[\lambda\frac{d}{d\lambda} + n\gamma(\lambda)\right] G_n(k_1 \ldots k_n, \lambda, g) = 0. \tag{2.36}$$

The solution of this equation is found without difficulty:

$$G_n(k_1 \ldots k_n, \lambda, g) = e^{-n\int_{\lambda_0}^{\lambda} \frac{d\lambda' \gamma(\lambda')}{\lambda'}} G_n(k_1 \ldots k_n, \lambda_0, g_0), \tag{2.37}$$

where $g_0 = g(\lambda_0)$, and λ_0 is a certain fixed normalization point, which we shall choose in the form $\lambda = \mathcal{N}\lambda_0$. Now we shall perform a scale transformation of the momenta: $k_i \to \mathcal{N}k_i$. From the equation (2.37) we obtain

$$G_n(\mathcal{N}k_1 \ldots \mathcal{N}k_n, \mathcal{N}\lambda_0, g) = e^{-n\int \frac{d\lambda' \gamma(\lambda')}{\lambda'}} G_n(\mathcal{N}k_1 \ldots \mathcal{N}k_n, \lambda_0, g). \tag{2.38}$$

If the dimension of the function G_n is d, then upon multiplication of all the momenta and the normalization point by $\mathcal{N}$, it is multiplied by $(\mathcal{N})^4$. Therefore we have

$$G_{(n)}(\mathcal{N}k_1 \ldots \mathcal{N}k_n, \mathcal{N}\lambda_0, g) = (\mathcal{N})^d G(k_1 \ldots k_n, \lambda_0, g). \tag{2.39}$$

Substituting this expression into (2.38), we obtain the law for transformation of the Green function in the case of homogeneous extension of all the momenta,

$$G_{(n)}(\mathcal{N}k_1, \ldots, \mathcal{N}k_n, \lambda_0, g_0) = \mathcal{N}^d e^{n\int_{\lambda_0}^{\lambda} \frac{d\lambda' \gamma(\lambda')}{\lambda'}} G_n(k_1, \ldots, k_n, \lambda_0, g(\mathcal{N}\lambda_0)). \tag{2.40}$$

One can see that the transformation law of the Green functions is not determined by naive dimensional arguments but includes an additional factor:

$$e^{n\int_{\lambda_0}^{\lambda} \frac{d\lambda' \gamma(\lambda')}{\lambda'}}. \tag{2.41}$$

This factor is due to the scaling transformation also altering the position of the normalization point. Hence arises the additional "anomalous" dimension of the Green functions. This factor is essentially of the same nature as the axial anomaly. Classical symmetry is violated by the renormalization procedure, which introduces into the theory a new dimensional parameter. If $g(\lambda)$ tends to a certain limit equal to g_∞, as $\lambda \to \infty$, while $\delta(g)$ tends toward $\delta(g_\infty)$, then up to a constant,

$$e^{n\int_{\lambda_0}^{\mathcal{N}\lambda} \frac{d\lambda' \gamma(g(\lambda'))}{\lambda'}} \to e^{n\gamma(g_\infty)\ln \mathcal{N}}. \tag{2.42}$$

$$\mathcal{N} \to \infty$$

Thus, the scale transformation of the Green functions is described by the formula

$$G_n(\mathcal{N}k_1 \ldots \mathcal{N}k_n, \lambda_0, g_0) = \mathcal{N}^{d+n\gamma_\infty} G_n(k_1 \ldots k_n, \lambda_0, g(\mathcal{N}\lambda_0)). \tag{2.43}$$

Extension of all the momenta by a factor of $\mathcal{N}$ leads to multiplication by a scale factor that takes into account the anomalous dimension and to substitution of the running constant $g(\mathcal{N}\lambda_0)$ for the coupling constant g_0.

In asymptotically free theories the scale behavior of Green functions as $\mathcal{N} \to \infty$ can be determined exactly. In this case no power corrections are present, but there does exist a logarithmic deviation from the scale transformation of free theory.

The analysis carried out above reveals that in asymptotically free theories the running constant $g(\mathcal{N})$ is small at high Euclidean momenta, and we can obtain reliable predictions applying perturbation theory.

Contrariwise, at $\mathcal{N} < 1$ the effective coupling constant increases. Of course, in this case the formula (2.27) obtained by means of the perturbation theory cannot be used. Nevertheless, if the β-function has no zeros at $g > 0$, then it follows from the equation (2.15) that

$$\bar{g}(\mathcal{N}, g) \to \infty, \qquad \mathcal{N} \to 0. \tag{2.44}$$

Such a behavior of the invariant charge would mean that with increasing distance the interaction strength would increase indefinitely, and consequently the particles would not be able to withdraw from each other to large distances.

The qualitative picture described above is realized in the hypothetical model of strong interactions known as "quantum chromodynamics." In this model hadrons are considered to be quark-bound states. Several types of quarks exist which differ from each other by the quantum number "flavor." Strangeness and charm are examples of "flavors." Each quark in its turn can exist in three varieties, differing in "color." Thus, the quarks are represented by the following matrix:

$$\begin{pmatrix} u_r, & u_y, & u_b \\ d_r, & d_y, & d_b \\ c_r, & c_y, & c_b \\ s_r, & s_y, & s_b \\ \cdot & \cdot & \cdot \end{pmatrix} \tag{2.45}$$

Here the indices r, y, b stand for the "colors" (red, yellow, blue) and the letters u, d, c, s indicate the various "flavors." The interaction between quarks is due to the exchange of colored Yang-Mills fields, that is, "gluons." The gauge group SU acts in the color space. The Yang-Mills fields form a color octet and are neutral with respect to flavors. The strong-interaction Lagrangian has the form

$$\mathcal{L} = \frac{1}{8}\mathrm{tr}\{\mathcal{F}_{\mu\nu}\mathcal{F}_{\mu\nu} + \bar{q}\{i\gamma_\mu[\partial_\mu - g\Gamma(\mathcal{A}_\mu)] - m\}g; \tag{2.46}$$

$$q = u, d, \ldots$$

The color SU_3^c symmetry is assumed to be exact.

The observed spectrum of hadrons is generated by (colorless) quark bound states that are singlets with respect to the group SU_3^c. In the approximation where all quarks have equal masses, the Lagrangian (2.46) is invariant under the transformations of the group SU^f acting in the space of flavors. Therefore, it is convenient to classify the hadron spectrum with respect to the group SU^f. However, owing to the great difference between the quark masses, the SU^f symmetry is strongly violated, and no degeneracy in the masses within the hadron multiplets is present.

From the point of view of theory, why quarks are not observed in a free state still remains the most unclear issue within this scheme. For explaining this fact the hypothesis of quark "confinement" is proposed. According to this hypothesis, owing to the special nature of interaction between quarks manifesting itself in the phenomenon of asymptotic freedom, colored objects, quarks, and gluons, are constantly combined into colorless complexes. Only these colorless complexes corresponding to real hadrons are observed experimentally. A number of simplified models, as well as computer calculations, testify in favor of this hypothesis; however, there exists no rigorous proof of it yet.

At present more far-reaching models unifying within the framework of a unique gauge theory all sorts of interactions are under intensive development. These models are based on the assumption that at superhigh energies exceeding 10^{17} GeV, the running constants of all interactions become equal and are combined into a single constant, while the dynamics of all interactions is determined by the Yang-Mills theory corresponding to a unique gaugegroup G (the most simple version of "Grand Unification" can be constructed on the basis of the $SU(5)$ group; however, this model seems to contradict the available experimental data). At energies characteristic of the Grand Unification the essential difference between leptons and baryons vanishes, and processes become possible involving nonconservation of the baryon number, for example, proton decay. At lower energies spontaneous symmetry breaking results in a reduction of the invariance group to $SU(3) \times SU(2) \times U(1)$, where $SU(3)$ represents the symmetry group of quantum chromodynamics; and $SU(2) \times U(1)$, of electroweak interactions. Finally, at energies of the order of 100 GeV the symmetry of electroweak interactions reduces to $U(1)$, the invariance group of electrodynamics. The Grand-Unification models yield a series of predictions interesting from an experimental point of view, but their detailed discussion is outside the scope of this book. We shall only point out that a consistent program for achieving such unification must also take the gravitational interaction into account, which will lead in a natural way to the consideration of supersymmetric theories exhibiting fermion-boson symmetry. Finally, during recent years the hypothesis has become widely acepted that quantum-field theories, such as chromodynamics and the theory of electroweak interactions, merely represent a low-energy limit of the fundamental theory describing the interaction of extended objects, namely, relativistic strings.

It is worth noting that although the models of relativistic strings are not described by field theory in the conventional sense, nevertheless, many of the methods described above find application in this field, too. This is due to relativistic strings representing still another example of a gauge-invariant system, and the same problems pertain to this system as the ones that arise in the Yang-Mills field theory and which were discussed in detail in this book.

Bibliography

Notes

As was pointed out in the preface, our book is not a textbook on quantum field theory. There exist many comprehensive manuals on quantum field theory among which we shall, first of all, name the classical monograph by N. N. Bogolubov and D. V. Shirkov [1]. We nevertheless consider the exposition, adopted in this book, of modern quantum field theory to be intrinsically consistent, in spite of the comparatively small volume of the book. Our description of quantum dynamics is based on the method of the path integral.

The application of this method to quantum-mechanical problems is expounded in the book by R. Feynman and A. Hibbs [2], and the monographs recently published by A. N. Vasil'ev [3] and V. N. Popov [4] are dedicated to the use of this method in the theory of systems with an infinite number of dgrees of freedom. The classical geometrical aspects of gauge fields have been considered in a monograph by N. P. Konoplyova and V. N. Popov [5], and their quantization and application to elementary-particle models have been briefly described in the book by J. Taylor [6].

As far as we know, the exposition proposed in the first edition of our book of the formalism of quantum field theory in terms of the path integral was the first attempt in this direction. Subsequently, the educational monographs by P. Ramon [7] and Zh. B. Zuber and K. Itzixson [8] were published, in which the path integral occupies significant space. We hope the corrections made in the second edition of our book have made it more consistent and will permit utilization of this book as a manual on modern quantum field theory.

Chapter 1

Gauge fields were first introduced in physics in the work of C. N. Yang and R. L. Mills [9] for fields carrying the interaction of isotopic spins. The natural generalization to the case of internal degrees of freedom of a more general nature is discussed, for example, in the publications [10, 11, 12].

Feynman was the first to draw attention to the specific character of the quantization of non-Abelian gauge fields [13]. His approach, based on the reconstruction of diagrams with loops by means of tree diagrams, was developed by De Witt [14], who formulated the final rules for quantization of gauge fields and gravity fields in [15]. An independent derivation of the rules of perturbation theory for these theories, based on the path integration, was obtained by L. D. Faddeev and V. N. Popov in [16] (see also [17]). The publications [18, 19, 20] are also dedicated to the construction of the perturbation theory for gauge fields. The hypothesis stated in Feynman's lectures [13] that the perturbation theory for gauge fields may be obtained in the limit $m \to 0$ of the theory of massive vector fields turned out to be incorrect [21, 22]. The first realistic unified interaction models, based on the Higgs mechanism [23, 24, 25], were formulated by S. Weinberg [26] and A. Salam [27].

In 1971, G. t'Hooft extended the quantization procedure for Yang-Mills fields to the case of theories with spontaneously broken symmetry [28]. In 1971-1972, in a series of publications by A. A. Slavnov [29, 30], J. Taylor [31], B. Lee and J. Zinn-Justin [32], and G. t'Hooft and M. Veltman [33], the methods were developed of invariant regularization and renormalization for the theory of gauge fields (including models with spontaneously broken symmetry), and thus the construction of the quantum theory of gauge fields in the framework of perturbation theory was completed.

From the viewpoint of differential geometry the classical Yang-Mills field represents a connection in the principal fiber-bundle space, the basis of which is the space-time manifold and a typical fiber of which is the internal symmetry group. The concept of connection, generalizing the Euclidean connection in the Riemann space, has been developed starting from the twenties in the work of many geometers, specifically, of H. Weyl and E. Cartan. Its modern formulation first appeared in the work of Ehresmann [34]. An excellent introduction to the theory of fiber bundles and connections may be found in the book by Lischnerovitch [35].

In the twenties, due to the success of the theory of general relativity, many attempts were made to geometrize the electromagnetic field. A correct view of this field as a part of the connection involved in the covariant derivative of the complex fields appeared in the work of H. Weyl [36] and V. A. Fock [37] on the formulation of the Dirac equation in the gravitational field. H. Weyl speaks directly about electrodynamics as the general relativity theory in the charge space.

The classical solutions of the equations of motion involving gauge fields have been a subject of intensive research during the last three years. We give references to some of the main publications in this field [38, 39, 40, 41], in which solutions for the vacuum and the soliton sector are studied.

Chapter 2

The path integral was first introduced for the formuation of quantum mechanics by Feynman. The history and main concepts may be found in the monograph [2]. The Feynman diagrams in the perturbation theory, first introduced in [41], were substantiated by means of the path integral in [43]. The monographs [3,4,7,8] contain a more up-to-date review of the applications of the path-integral method in quantum physics. The interpretation of the path-integral method in the present book follows the lectures of one of the authors [44]. The introduction of the path integral in quantum mechanics by a formula similar to (1.12) of Chapter 2 is adopted from the work of Tobocman [45]. The holomorphic representation of quantum mechanics is from the work of V. A. Fock; it appeared under the name of coherent states in quantum optics. Its mathematical formulation may be found in the monograph of F. A. Berezin [46]. There also the first rigorous exposition of the integration over anti-commuting variables is given.

The boundary conditions in the path integral were considered by O. I. Zav'yalov [47] and A. N. Vasil'ev [3].

The Green functions were introduced in the quantum field theory by J. Schwinger [48]. The idea of reducing the calculation of the S-matrix to the problem of the calculation of the S-matrix for scattering by an external source also belongs to Schwinger [49].

Representation of the S-matrix in the form of a path integral with asymptotic conditions at large times was discussed in Feynman's lectures [13].

A closed form for the S-matrix leading naturally to expansion in loops was given in [50] (see also the monograph by A. N. Vasil'ev [13]).

The introduction of the path integral in terms of the Gaussian functional was proposed in the first edition of the monograph [1]. The exposition in the present book of the axiomatics of the path integral by means of the Gaussian functional follows the work of one of the authors [51]. A similar approach was developed also in [52].

Chapter 3

The generalized Hamiltonian dynamics was first introduced by Dirac [53] (see also his lectures [54]). The Hamiltonian formulation of gauge theories in the Coulomb gauge was investigated by J. Schwinger [55]. The general formulation of the path integral in the generalized Hamiltonian form was given by one of the authors [56].

The exposition of quantization in the Hamilton gauge follows [57].

The change-of-variables method in the path integral to pass from one gauge to another was proposed in [16]. Its geometrical interpretation in terms of various parametrizations of gauge-equivalent classes is discussed in [56, 58]. The generalized gauges were first considered accurately in [14] (see also [19, 56]). The method of transition to the α-gauge, described in the present book, is adopted from the work of G. t'Hooft [28].

Chapter 4

The renormalization theory goes back to the idea of H. A. Kramers [59] and H. A. Bethe [60]. A number of authors, including R. P. Feynman, J. Schwinger, F. Dyson, A. Salam, and others, took part in its development. The complicated, mathematically rigorous renormalization theory (the theory of R-operation) was first constructed in the work of N. N. Bogolubov and O. S. Parasyuk [61]. An excellent exposition of the R-operation may be found in the monographs [1, 62], where a detailed bibliography on the renormalization theory is also presented.

Regularization by means of higher covariant derivatives was first propounded in the work of A. A. Slavnov [63] and then applied to the Yang-Mills theory in [28, 30]. The additional regularization of one-loop diagrams, described in Section 4.4, was constructed in [64].

The dimensional regularization was proopsed in the work by G. t'Hooft and M. Veltman [31], C. G. Bollini and J. G. Giabidgi [65], and J. F. Ashmore [66].

Gauge fields on a lattice were first introduced by K. Wilson [67]. Detailed exposition of the formalism of gauge fields on a lattice and of its application to numerical computations can be found in the monograph by M. Creutz [68].

The identities relating two- and three-point Green functions in quantum electrodynamics were first obtained by J. C. Ward [69]. The generalized relations, connecting any Green function with the function containing one external photon line less, were obtained by E. S. Fradkin [70] and Y. Takahashi [71]. The electrodynamical Ward identities are not generalizable directly to the case of non-Abelian gauge fields. In the non-Abelian theory their role is played by the so-called generalized Ward identities, first obtained by A. A. Slavnov [29] and J. C. Taylor [31]. Their derivation, given in the present book, follows [72]. An alternative derivation based on the use of the invariance of the effective Lagrangian with respect to some transformation with anticommuting parameters (supertransformation) was proposed by C. Becchi, A. Rouet, and R. Stora [73]. In the literature the generalized Ward identities for single-particle irreducible Green functions, obtained by B. W. Lee [74] (see also [52]), are also used. The structure of the renormalized action was investigated in [29, 31, 32, 33, 75], as a result of which there appeared a proof of gauge-invariance and unitarity of the renormalized S-matrix. Another approach to the renormalization of gauge theories, based on the use of the formalism of the Zimmerman normal products, was developed by C. Becchi, A. Rouet, and R. Stora [73, 76].

The dependence of the renormalization constants and the Green functions on the choice of the gauge condition is discussed in detail in the work by R. Callosh and I. Tyutin [77].

Covariant quantization based on BRS-invariance was proposed in [78]. The description of the classification of representations of the BRS-algebra presented in this book follows [79].

The anomalous Ward identities were first studied by S. L. Adler [80] and by S. Bell and R. Jackiw [81]. Their role in the problem of the renormalization of gauge theories is discussed in [82–85].

The condition of the self-consistence for non-Abelian anomalies was obtained by J. Wess and B. Zumino [86]. The explicit expression for a non-Abelian anomaly given in this book was taken from the work of W. Bardeen [87]. The topological approach to the classification and calculation of anomalies was proposed in the works by B. Zumino [88], and R. Stora [89]; see also [90]. The manifestation of anomalies in the Hamiltonian gauge $A_0 = 0$ changing the commutation relations for the Gauss law (2.7) of Chapter 3 is discussed in [91].

The classification of anomalous interactions present in this book follows the work by H. Georgi and S. Glashow [85].

Chapter 5

The unified model of weak and electromagnetic interactions was constructed by S. Weinberg [26], A. Salam [27], and S. Glashow [92].

The mechanism of suppression of the strangeness-changing neutral currents by the introduction of a new "charmed" quark was propounded by S. Glashow, J. Illiopoulos and L. Maiani [93].

The hypothesis of the existence of an additional degree of freedom of the quarks, which became known as "color," was first proposed in the work of N. N. Bogolubov, B. V. Struminsky, and A. N. Tavkhelidze [94], and of M. Y. Han and Y. Nambu [95] for the explanation of the hadron statistics. In the work of J. Pati and A. Salam [96], of H. Fritzsch, M. Gell-Mann, and H. Leutwyller [97], and S. Weinberg [98] it was first suggested that strong interactions take place owing to the exchange of Yang-Mills mesons interacting with the color degrees of freedom. The corresponding model was named quantum chromodynamics.

We refer the reader interested in applications of gauge fields to the phenomenology of elementary particles, including issues related to the Grand Unification, to the monographs by L. B. Okun [111] K. Huang [112] and I. Nambu [113].

The group features of the renormalization transformations were first drawn to attention by E. C. G. Stueckelberg and A. Peterman [99]. The group of multiplicative renormalizations in quantum electrodynamics was used by M. Gell-Mann and F. Low [100] for the investigation of the ultraviolet asymptotics of the Green functions. The general theory of the renormalization group was constructed in the work of N. N. Bogolubov and D. V. Shirkov [101, 102]. A detailed exposition of this theory may be found in the monograph [1]. The differential equations of the renormalization group were investigated by L. V. Ovsyannikov [103]. Analogous equations were obtained in the frame of quantum field theory by C. Callan [104] and K. Symanzik [105]. The asymptotic freedom of the Yang-Mills fields was discovered by G. t'Hooft [106], by D. Gross and F. Wilczek [107], and by H. D. Politzer [108]. The hypothesis of quark confinement was discussed in [97, 98, 109, 100].

A

Anomalous Commutator of the Gauss Law

In Section 11 of Chapter 4 we mentioned a possible change of the quantization procedure of gauge fields in models involving anomalies. Here we shall present this point of view in greater detail.

Consider the example of a model describing the interaction of chiral fermions realizing a representation of the non-Abelian group Ω with the Yang-Mills field. The action in such a model is given by the expression

$$S = S_{YM} + \int dx\, i\bar{R}(x)\gamma_\mu[\partial_\mu - \Gamma(\mathcal{A}_\mu)]R(x), \tag{A.1}$$

where S_{YM} stands for the action of the Yang-Mills field in vacuum,

$$R = \frac{1}{2}(1-\gamma_5)\psi \tag{A.2}$$

$\Gamma(\mathcal{A}_\mu)$, as usual, denotes the representation of the matrix $\mathcal{A}_\mu$, corresponding to the given representation of the group Ω. Below we shall for simplicity set

$$\Gamma(\mathcal{A}_\mu) = A^a_\mu \lambda^a \equiv A_\mu, \tag{A.3}$$

where

$$[\lambda^a, \lambda^b] = t^{abc}\lambda^c, \qquad \operatorname{tr}(\lambda^a\lambda^b) = \frac{1}{2}\delta^{ab}. \tag{A.4}$$

In the Hamilton gauge $A_0 = 0$ the relationship between the constraint and the Gauss law has the form

$$C^a(x) = \partial_k F^a_{k0} - t^{abc}A^b_k F^c_{0k} + i\bar{R}\lambda^a\gamma_0 R \tag{A.5}$$

(compare with (4.6) of Chapter 3).

In classical theory the constraits C^a satisfy the commutation relations

$$\{C^a(x), C^b(y)\} = t^{abc}\delta(x-y)c^c(x) \tag{A.6}$$

As we saw in Chapter 3, these relations allow interpretation of the condition $C^a(x) = 0$ as the constraint of the first class, which in quantum theory becomes the condition imposed on admissible state vectors,

$$C^a(x)\Psi = 0, \tag{A.7}$$

where ψ is an admissible physical vector. Below we shall show that in the presence of anomalies this condition may be violated, and the determination of physical states requires special discussion. To demonstrate this, we consider the functional

$$Z(A) = \int \exp\{iS(A, \bar{\psi}, \psi)\} \prod_x d\bar{\psi}(x) d\psi(x), \tag{A.8}$$

in which the Yang-Mills fields are considered external. Proceeding precisely as in Chapter 4 we can readily obtain for this functional the anomalous Ward identity, which can be written in the form

$$\left(\nabla_\mu \frac{\delta}{\delta A_\mu(x)}\right)^a Z(A) = iG^a(x)Z(A), \tag{A.9}$$

where $G(x)$ is the anomaly,

$$G^a(x) = \frac{i}{24\pi^2}\varepsilon^{\mu\nu\lambda\sigma}\partial_\mu \operatorname{tr}\left[\lambda^a\left(A_\nu\partial_\lambda A_\sigma - \frac{1}{2}A_\nu A_\lambda A_\sigma\right)\right]. \tag{A.10}$$

Note that the concrete form of the anomaly may depend on the utilized regularization: Various regularizations can lead to results differing by the derivatives of local counterterms. We make use of the expression encountered most often in the literature.

In the language of the path integral an anomaly may be interpreted as follows. As can be seen from the formula (A.9), the anomaly represents a variation of the functional $\ln Z$ under an infinitesimal gauge transformation of the fields A_μ. On the other hand, from the integral representation (A.8) for $Z(A)$, we can see that if, at the same time as the gauge transformation of the fields A_μ, we perform a compensating change of variables,

$$\delta\psi = u\psi, \quad \delta\bar{\psi} = -\bar{\psi}u, \quad \omega = e^u, \tag{A.11}$$

then the variation $Z(A)$ can be represented in the form

$$\delta Z(A) = \int \exp\{iS\} \prod_x J d\bar{\psi} d\psi, \tag{A.12}$$

where $J = 1 + \delta J$ is the Jacobian of the transformation (A.11). Hence it follows that

$$\delta \ln Z(A) = \ln J. \tag{A.13}$$

Comparing this expression with the definition of an anomaly (D.9) we obtain

$$\delta J = -i \int dx G^a(x) u^a(x). \tag{A.14}$$

From this formula we can see that an anomaly represents the variation of integration measure $\prod_x d\bar{\psi}d\psi$ under an infinitesimal gauge transformation of the fields ψ. From a formal point of view this measure is invariant with respect to the transformations (A.11). However, as we know, to provide for the integral (A.8) having a rigorous meaning, it is necessary to introduce regularization (for instance, the Pauli-Villars regularization). It is convenient to consider all the regularizing factors in terms of the integration measure. Under such convention the measure becomes dependent on the fields A_μ, and its variation is related to the anomaly by the formula (A.14).

For finite transformations the corresponding formula has the form

$$\prod_x d\bar{\psi}^\omega d\psi^\omega = e^{i\alpha(A,\omega)} \prod_x d\bar{\psi} d\psi. \tag{A.15}$$

The function $\alpha(A,\omega)$ is determined from the differential equation

$$\delta_\omega \alpha(A,\omega) = G(A^\omega), \tag{A.16}$$

which can be integrated explicitly. Further we shall need the expression for the phase $\alpha(x)$ with an accuracy up to second-order terms in u. This expression has the form

$$\alpha = \frac{i}{48\pi^2} \varepsilon^{\mu\nu\lambda\sigma} \mathrm{tr} \Big[(A_\mu \partial_\nu A_\lambda + \partial_\nu A_\mu A_\lambda - A_\mu A_\nu A_\lambda) \times \left(\partial_\sigma u - \frac{1}{2}[\partial_\sigma u, u] \right) - \frac{1}{2} A_\mu \partial_\nu u A_\lambda \partial_\sigma u \Big] + O(u^3). \tag{A.17}$$

Now let us pass to the calculation of the commutator for the constraints. In the operator formalism the matrix element of the commutator of two field operators $A(x)$ and $B(x)$ can be expressed through the matrix element of the chronological product of the same operators with the aid of the formula

$$\lim_{q_0 \to \infty} q_0 \int dt' e^{iq_0(t'-t)} \langle \alpha | T A(x,t') B(y,t) | \beta \rangle = i \langle \alpha | [A(x,t), B(y,t)] | \beta \rangle, \tag{A.18}$$

which is called the Johnson-Bjorken-Low formula. In the path-integral formalism the corresponding formula has the form

$$\langle \alpha | [A(x,x_0), B(y,x_0)] | \beta \rangle = -i \lim_{q_0 \to \infty} q_0 \int e^{iq_0(y_0 - x_0)} \times \int_{A_\mu \to A_{\mu\binom{\mathrm{in}}{\mathrm{out}}},\, \psi \to \psi_{\binom{\mathrm{in}}{\mathrm{out}}}} \exp\{iS\}(A(x), B(y) \prod_x d\bar{\psi} d\psi dA_\mu \tag{A.19}$$

where the asymptotic fields $A_{\mu\binom{\mathrm{in}}{\mathrm{out}}}$ and $\psi_{\binom{\mathrm{in}}{\mathrm{out}}}$ correspond to the states α and β, respectively.

To obtain the expression for the commutator of the constraints, it is sufficient to write the Ward identity for the chronological product of the constraints. We recall that a constraint $C^a(x)$ is the coefficient of A_0 in the action (A.1) written in the first-order formalism (see Section 4). Therefore for our purposes it is natural to take advantage of the device repeatedly applied in Chapter 4.

Now we shall perform the following gauge change of variables:

$$\psi \to \psi^\omega, \quad \bar{\psi} \to \bar{\psi}^\omega, \quad A_\mu \to A_\mu^\omega, \quad E \to E^\omega \tag{A.20}$$

in the generating functional for the S-matrix in the Hamiltonian gauge

$$Z_s = \int\limits_{A_\mu \to A_\mu\binom{\text{in}}{\text{out}}, \psi \to \psi\binom{\text{in}}{\text{out}}} \exp\{iS\}(E, A, \bar{\psi}, \psi)\delta(A_0) \times \prod_x dA_\mu dE d\bar{\psi} d\psi. \tag{A.21}$$

As a result of this transformation, the integral (A.21) assumes the form

$$Z_s = \int\limits_{A_\mu \to A_\mu\binom{\text{in}}{\text{out}}, \psi \to \psi\binom{\text{in}}{\text{out}}} \exp\{iS\}\delta(A_0)\exp\left\{i\int(\partial_\mu\omega^{-1}\omega)^a \right.$$
$$\left. \times C^a(x)dx + i\alpha(A,\omega)|_{A_0=\partial_0\omega^{-1}\omega}\right\} \prod_x dA_\mu dE d\bar{\psi} d\psi. \tag{A.22}$$

In writing this expression we explicitly utilized the formula (A.15) for transforming the functional measure.

To obtain the Ward identities of interest, we equate to zero the second-order terms in u in the formula (A.22). By construction, the integral (A.22) is independent of u; therefore all the terms of any order in u should become, independently of each other, equal to zero. Substituting into the expression (A.17) for $\alpha(A,\omega)A_0 = \partial_0\omega^{-1}\omega$ and writing out explicitly the second-order terms in u, we obtain the relation

$$-\frac{1}{2}\int dx\, dy \langle TC^a(x)C^b(y)\rangle \partial_0 u^a(x)\partial_0 u^b(y) - \frac{i}{2}\int dx t^{abc} u^a(x)\partial_0 u^b(x)\langle C^c(x)\rangle$$
$$= -\frac{1}{48\pi^2}\int dx \varepsilon^{ijk} \operatorname{tr}[\partial_l A_j, (\partial_k u \partial_0 u + \partial_0 u \partial_k u)] + \ldots, \tag{A.23}$$

where the dots indicate the terms containing no derivatives of u with respect to time, which are of no significance in the Bjorken-Johnson-Low limit. The functions C^a are determined by the formula

$$C^a = C^a - \frac{i}{48\pi^2}\varepsilon^{ijk}\operatorname{tr}[\lambda^a(A_i\partial_j A_k + \partial_j A_k A_i - A_i A_j A_k)]. \tag{A.24}$$

In deriving this formula the following change of variables was made:

$$E_i^a \to E_i^a + \frac{i}{48\pi^s}\varepsilon_{ijk}\operatorname{tr}[\lambda^a\{A_j, \partial_k\omega^{-1}\omega\}]. \tag{A.25}$$

The relation (A.23) represents an identity of the form

$$\int K^{ab}(x,y)\, u^a(x)\, u^b(y)\, dx\, dy = 0, \tag{A.26}$$

where $K^{ab}(X,y)$ is the kernel that is symmetric with respect to the transposition $(x, a \leftrightarrow y, b)$. The Fourier transform of the first term differs from the matrix element of interest of the chronological product of the constraints by the factor $p_0 q_0$. Therefore, for obtaining the equal-time commutator of the constraints it is necessary, in accordance with the formula (A.18), to act on $K^{ab}(x,y)$ by the operator

$$\lim_{(p_0-q_0)\to\infty} \frac{p_0-q_0}{p_0 Q_0} \int dx_0\, dy_0 e^{ip_0x_0+iq_0y_0}, \tag{A.27}$$

Application of this operator to the second and third terms K^{ab} removes from them the time derivatives. This fact follows from the trivial equality

$$\frac{p_0-q_0}{p_0q_0} = \frac{1}{q_0} - \frac{1}{p_0}.$$

The terms containing no time derivatives give no contribution in this limit. As a result, we obtain the relation

$$[C^a(x), C^b(y)] = it^{abc}C^c(y)\delta(x-y) - \frac{1}{24\pi^2}\varepsilon_{ijk}\mathrm{tr}[\partial_i A_j\{\lambda^a, \lambda^b\}]\partial_k\partial(x-y). \tag{A.28}$$

If we go back from the operators C^a to the initial constraints C^a, then for the Gauss law we obtain the anomalous commutator

$$\begin{aligned}[C^a(x), C^b(y)] &= it^{abc}C^c(y)\partial(x-y) \\ &\quad -\frac{1}{48\pi^2}\varepsilon_{ijk}\mathrm{tr}[[\lambda^a,\lambda^b](A_i\partial_j A_k + \partial_j A_k A_i - A_i A_j A_k) \\ &\quad -\partial_j A_i \lambda^a A_k \lambda^b + \partial_i A_j \lambda^b A_k \lambda^a]\delta(x-y).\end{aligned} \tag{A.29}$$

The formula (A.29) shows that in the anomalous model the constraints, or at least some of them, are constraints of the second class; i.e., their commutator does not vanish on the surface of the constraints. This means that the condition (A.7) singling out physical state vectors is, in this case, not self-consistent. Therefore for theories involving anomalies the quantization procedure described in Chapter 3 must be reconsidered. The indicated inconsistency represents one more manifestation of the contradiction, mentioned in Section 11 of Chapter 4, that arises in anomalous models when one attempts to restrict the physical spectrum to transverse states. This can be considered an indication that anomalous models describe interactions involving a higher number of degrees of freedom than the corresponding classical theories. There exist arguments in favor of the fact that, when the space of asymptotic states is chosen correctly and noncummutativity is taken into account properly, then it is possible to construct a consistent quantization procedure for theories containing anomalies. However, since this issue is not yet quite clear, we shall not discuss it here in greater detail, anu we refer the reader to journals. We shall only point out that to carry out such a program requires application of methods not related to expansion in the coupling constant.

A.1 Bibliography

The above exposition, in the main, follows [114, 115]. The interpretation of an anomaly as a change of the integration measure in the path integral was first proposed by K. Fujikawa [116, 117]. A good description of the calculation of the equal-time commutator with the aid of the Bjorken-Johnson-Low procedure is given in [118]. The possibility of the appearance of an anomalous commutator was discussed for the first time in [119]. The anomalous commutator in the form (A.29) was derived in [120, 121] by performing summation of perturbation-theory diagrams.

References

1. Bogolubov, N. N., Shirkov, D. V., *An Introduction to the Theory of Quantized Fields*. Moscow, Nauka, 1976.

2. Feynman, R. P., Hibbs, A. R., *Quantum Mechanics and Path Integrals*. New York, McGraw-Hill, 1965.

3. Vasil'ev, A. N., *Functional Methods in Quantum Field Theory and Statistics*. Leningrad, Leningrad University Press, 1976.

4. Popov, V. N., *Path Integrals in Quantum Field Theory and Statistical Physics*. Moscow, Atomizdat, 1976.

5. Konoplyova, N. P., Popov, V. N., *Gauge Fields*. Moscow, Atomizdat, 1972.

6. Taylor, J. C., *Gauge Theories of Weak Interactions*. Cambridge, Cambridge University Press, 1976.

7. Ramond, P., *Field Theory: A Modern Primer*. London, Benjamin/Cummings, 1981.

8. Itzykson, C., Zuber J-B., *Quantum Field Theory*. New York, McGraw-Hill, 1980.

9. Yang, C. N., Mills, R. L., *Phys. Rev.*, ***96***, 191, 1954.

10. Utiyama, R., *Phys. Rev.*, **101**, 1597, 1956.

11. Salam, A., Ward, J. C., *Nuovo Cimento*, **XI**, 568, 1959.

12. Glashow, S. L., Gell-Mann, M., *Ann. of Phys.*, **15**, 437, 1961.

13. Feynman, R., *Acta Phys. Polonica*, **24**, 697, 1963.

14. De Witt, B., *Phys. Rev. Lett.*, **12**, 742, 1964.

15. De Witt, B., *Phys. Rev.*, **160**, 1113; 1195, 1967.

16. Faddeev, L. D., Popov, V. N., *Phys. Lett.*, **B25**, 30, 1967.

17. Faddeev, L. D., Popov, V. N., Preprint, Institute for Theoretical Physics, Ukrainian Acad. Sci., Kiev, 1967.

18. Mandelstam, S., *Phys. Rev.*, **175**, 1580, 1968.

19. Fradkin, E. S., Tyutin, I. V., *Phys. Lett.*, **B30**, 562, 1969.

20. Vainstein, A. I., Khriplovich, I. B., *Yadernaya Fizika*, **13**, 198, 1971.

21. Boulware, A., *Ann. Phys.*, **56**, 140, 1970.

22. Slavnov, A. A., Faddeev, L. D., *TMP*,[1] 3, 18, 1970.

23. Higgs, P. W., *Phys. Lett.*, 12, 132, 1964.

24. Englert, F., Brout, R., *Phys. Rev. Lett.*, **13**, 321, 1964.

25. Kibble, T. W. B., *Phys. Rev.*, **155**, 1554, 1967.

26. Weinberg, S., *Phys. Rev. Lett.*, **19**, 1264, 1967.

27. Salam, A., *Elementary Particle Theory*. N. Svartholm (ed.). Stockholm, Almquist, Forlag AB, 1968.

28. t'Hooft, G., *Nucl. Phys.* **B35**, 167, 1971.

29. Slavnov, A. A., *TMP*, **10**, 99, 1972.

30. Slavnov, A. A., *TMP*, **13**, 174, 1972.

31. Taylor, J. C., *Nucl. Phys.*, **B33**, 436, 1971.

32. Lee, B. W., Zinn-Justin, J., *Phys. Rev.*, *D*, **5**, 3137, 1972.

33. t'Hooft, G. Veltman, M., *Nucl. Phys.*, **B44**, 189; **B50**, 318, 1972.

34. Ehresmann, *Coll.top. Bruxelles*, p. 29, 1950.

35. Lischnerovich, A., *Theory of Connections as a Whole and Holonomy Groups*. Moscow, Inostrannaya Literatura, 1960 (Russian translation).

36. Weyl, H., *Z.f. Phys.*, **56**, 330, 1929.

37. Fock, V., *Journ. de Physique*, **10**, 392, 1929.

38. Polyakov, A. M., *Lett. to JETP*, **20**, 430, 1974.

1. Teoreticheskay'a i Matematicheskay's Fizika.

39. t'Hooft, G., *Nucl. Phys.*, **B79**, 2761, 1974.

40. Faddeev, L. D. Preprint MPI-RAE/Pth München, 1974.

41. Belavin, A. A., Polyakov, A. N., Tyupkin, Y., Schwarz, A. S., *Phys. Lett.* **B59**, 85, 1975.

42. Feynman, R. P., *Rev. Mod. Phys.*, **20**, 367, 1948.

43. Feynman, R. P., *Phys. Rev.*, **80**, 440, 1950.

44. Faddeev, L. D., Les Houches Lecture, Session 20. North-Holland, 1976.

45. Tobocman, W., *Nuovo Cimento*, **3**, 1213, 1956.

46. Berezin, F. A., *The Method of Secondary Quantization*. Moscow, Nauka, 1965.

47. Zav'yalov, O. I., Thesis, Math. Inst. Acad. Sci., Moscow, 1970.

48. Schwinger, J., *Phys. Rev.*, **75**, 651, 1949.

49. Schwinger, J., *Proc. Nat. Acad. Sci.*, **37**, 452, 1951.

50. Aref'eva, I. A., Faddeev, L. D., Slavnov, A. A., *TMP*, *21*, 311, 1974.

51. Slavnov, A. A., *TMP*, **22**, 177, 1975.

52. Zinn-Justin, J., *Lecture Notes in Physics*, Vol. 37, Berlin, Springer-Verlag.

53. Dirac, P.A.M., *Proc. Roy. Sci.*, **A246**, 326, 1958.

54. Dirac, P.A.M., Lectures on Quantum Mechanics, New York, Yeshiva University, 1964.

55. Schwinger, J., *Phys. Rev.*, **125**, 1043, 1962; **127**, 324, 1962.

56. Faddeev, L. D., *TMP*, *1*, **3**, 1969.

57. Slavnov, A. A., Frolov, S. A., *TMP*, *68*, 360, 1986.

58. Popov, V. N., Faddeev, L. D., *Uspekhi Fiz. Nauk*, **111**, 427, 1973.

59. Kramers, H. A., *Rapports du* 8[e] *Conseil Solvay*, Bruxelles, 1950.

60. Bethe, H. A., *Phys. Rev.*, **72**, 339, 1947.

61. Bogolubov, N. N., Parasyuk, O. S., *Doklady Akad. Nauk USSR*, **55**, 149, 1955; **100**, 429, 1955, *Acta Math.* **97**, 227, 1957.

62. Zav'yalov, O. I., *Renormalized Feynman Diagrams*. Moscow, Nauka, 1979 (in Russian).

63. Slavnov, A. A., *Nucl. Phys.*, **B31**, 301, 1971.

64. Slavnov, A. A., *TMP*, **33**, 210, 1979.

65. Bollini, C. G., Giabidgi, J. T., *Phys. Lett.*, **B40**, 566, 1972.

66. Ashmore, J. F., *Nuovo Cimento Lett.*, **4**, 289, 1972.

67. Wilson, K., *Phys. Rev., D14*, 2445 1974.

68. Creutz, M., *Quarks, Gluons and Lattices*, Cambridge, Cambridge University Press, 1985.

69. Ward, J. C., *Phys. Rev.*, 77, 2931, 1950.

70. Fradkin, E. S., *JETP*, **29**, 288, 1955.

71. Takahashi, Y., *Nuovo Cimento*, **6**, 370, 1957.

72. Slavnov, A. A., *Nucl. Phys.*, **B97**, 155, 1975.

73. Becchi, C., Rouet, A., Stora, R., *Comm. Math. Phys.*, **42**, 127, 1975.

74. Lee, B. W., *Phys. Lett.*, **B46**, 214, 1974; *Phys. Rev.*, **9**, 933, 1974.

75. Slavnov, A. A., *Sov. J. of Particles and Nuclei*, **5**, 303 1975.

76. Becchi, C., Rouet, A., Stora, R., *Renormalization Theory* (G. Velo,; A. S. Wightman, eds.). D. Reidel Publ. Co., 1976; *Ann Phys.*, **98**, 287, 1976.

77. Kallosh, R., Tyutin, I., *Yadernaya Fizika*, **17**, 190, 1973.

78. Kugo, T., Ojima I., *Suppl. Progr. Theor. Phys.*, **N66**, 1979.

79. Nishijima, K., *Nucl. Phys., B238*, 601, 1984.

80. Adler, S. L., *Phys. Rev.*, **177**, 2426, 1969.

81. Bell, S., Jackiw, R., *Nuovo Cimento*, **A60**, 47, 1969.

82. Slavnov, A. A., *TMP*, **7**, 13, 1971.

83. Gross, D. J., Jackiw, R., *Phys. Rev. D*, **6**, 477, 1972.

84. Bouchiat, C., Illiopoulos, J., Meyer, P., *Phys. Lett.*, **B38**, 519, 1972.

85. Georgi H., Glashow, S. L., *Phys. Rev. D*, **6**, 429, 1977.

86. Wess, J., Zumino, B., *Phys. Lett., B37*, 95, 1971.

87. Bardeen, W. A., *Phys. Rev., 184*, 1848, 1969.

88. Zumino, B. In: *Relativity, Groups and Topology*. Amsterdam, North-Holland, 1983, p. 86.

89. Stora, R. Ibid., p. 34.

90. Reyman, A. G., Semyenov-Tyan'shanskij, M. A., Faddeev, L. D., *Functional Analysis, 18*, 64, 1984.

91. Faddeev, L. D., *Phys. Lett., B145*, 81, 1984.

92. Glashow, S. L., *Nucl. Phys., 22*, 579, 1961.

93. Glashow, S. L., Illiopoulos, J., Maiani, L., *Phys. Rev., D*, **2**, 185, 1970.

94. Bogolubov, N. N., Struminsky, B. V., Tavkhelidze, A. N., JINR Preprint D-1968, 1965.

95. Hann, M.Y., Nambu, Y., *Phys. Rev.*, **139**, B1006, 1965.

96. Pati, J., Salam A., *Phys. Rev. D*, **8**, 1240, 1973.

97. Fritzsch, H., Gell-Mann, M., Leutwyller, H., *Phys. Lett.*, **B47**, 365, 1973.

98. Weinberg, S., *Phys. Rev. Lett.*, **31**, 494, 1973.

99. Stueckelberg, E.C.G., Peterman, A., *Helv. Phys. Acta*, **26**, 499, 1953.

100. Gell-Mann, M., Low F., *Phys. Rev.*, **95**, 1300, 1954.

101. Bogolubov, N. N., Shirkov, D. V., *Doklady Akad. Nauk USSR*, **103**, 203, 391, 1955.

102. Bogolubov, N. N., Shirkov, D. V., *Nuovo Cimento*, **3**, 845, 1956.

103. Ovsyannikov, L. V., *Doklady Akad. Nauk*, **109**, 112, 1956.

104. Callan, C., *Phys. Rev. D*, **2**, 2, 1542, 1970.

105. Symanzik, K., *Comm. Math. Phys.*, **18**, 227, 1970.

106. t'Hooft, G., Report at the Conference on Lagrangian Field Theories, Marseille, 1972.

107. Gross, D., Wilczek, F., *Phys. Rev. D*, **8**, 3633, 1973.

108. Politzer, H. D., *Phys. Reports*, **C14**, 129, 1974.

109. Polyakov, A. M., *Phys. Lett.*, **B59**, 82, 1975.

110. Wilson, K., Erice Lectures. CNLS-321, 1975.

111. Okun, L. B., *Leptons and Quarks*. Moscow, Nauka, 1980.

112. Huang, K., *Quarks, Leptons, and Gauge Fields*, Singapore, World Scientific, 1982.

113. Nambu, I., *Quarks*, Moscow, Mir, 1985. (Russian translation from Japanese).

Supplementary References

114. Faddeev, L. D., Shatashvili, S. L., *Phys. Lett.*, ***167B***, 225, 1986.

115. Alekseev, Yu. A., Madajchik, Ya., Faddeev, L. D., Shatashvili, S. L., Preprint R-7-87, Leningrad, LOMI, 1987.

116. Fujikawa, K., *Phys. Rev.*, ***D21***, 2848, 1980.

117. Fujikawa, K., *Phys. Lett.*, ***171B***, 424, 1986.

118. Treiman, S., Jackiw, R., Witten, E., *Current Algebra and Anomalies.* Singapore, World Scientific, 1985.

119. Faddeev, L. D., *Phys. Lett.*, ***145B***, 82, 1984.

120. Jo, S.-G., *Phys. Lett.*, ***163B***, 353, 1985.

121. Kobayashi, M., Seo, K., Sugamoto, A., *Nucl. Phys.*, ***B273***, 607, 1986.

Notation

(x, y) are points in the Minkowski space, and $(\vec{x}, t), (\vec{y}, s)$ or $(\vec{x}, x_0), (\vec{y}, y_0)$ are their space and time components, respectively.

The metric tensor $g^{\mu\nu}$ has the form diag $(1, -1, -1, -1)$.

All vectors are assumed to be covariant; $ab = a_\mu b_\mu = a_0 b_0 - (\vec{a}, \vec{b})$ is the scalar product of the four-dimensional vectors a, b with the components $a_\mu b_\mu$; the components of four-dimensional vectors are labeled by Greek letters, and of three-dimensional vectors by Latin letters.

The constants $\hbar$ and c, unless otherwise stated, are taken to be equal to unity.

Milton Keynes UK
Ingram Content Group UK Ltd.
UKHW020028071024
449327UK00032B/2976

9 780201 406344